MATH ADVENTURES WITH PYTHON

MATH ADVENTURES WITH PYTHON

AN ILLUSTRATED GUIDE TO EXPLORING MATH WITH CODE

BY PETER FARRELL

San Francisco

Printed in USA

First printing

22 21 20 19 18 1 2 3 4 5 6 7 8 9

ISBN-10: 1-59327-867-5
ISBN-13: 978-1-59327-867-0

Publisher: William Pollock
Production Editor: Meg Sneeringer
Cover Illustration: Josh Ellingson
Developmental Editor: Annie Choi
Technical Reviewer: Patrick Gaunt
Copyeditor: Barton D. Reed
Compositors: David Van Ness and Meg Sneeringer
Proofreader: James Fraleigh

The following images are reproduced with permission:
Figure 10-2 by Acadac mixed from originals made by Avsa (*https://commons.wikimedia.org/wiki/File:Britain-fractal-coastline-100km.png#/media/File:Britain-fractalcoastline-combined.jpg*; CC-BY-SA-3.0);
Figure 11-19 by Fabienne Serriere, *https://knityak.com/*.

For information on distribution, translations, or bulk sales, please contact No Starch Press, Inc. directly:
No Starch Press, Inc.
245 8th Street, San Francisco, CA 94103
phone: 1.415.863.9900; info@nostarch.com
www.nostarch.com

A catalog record of this book is available from the Library of Congress.

This book is dedicated to all my students,
from whom I've learned so much.

ABOUT THE AUTHOR

Peter Farrell was a math teacher for eight years, starting first as a Peace Corps volunteer in Kenya. He then worked as a computer science teacher for three years. After reading Seymour Papert's *Mindstorms* and being introduced to Python by a student, he was inspired to bring programming into math class. He is passionate about using computers to make learning math more relevant, fun, and challenging.

ABOUT THE TECHNICAL REVIEWER

Paddy Gaunt graduated in engineering within weeks of the birth of the IBM PC and its associated MS DOS. Much of the rest of his career has revolved around implementing mathematical or technical concepts in practical software. Recently, he reformed links with Cambridge University (UK) when he became lead developer of pi3d, a python module for 3D graphics initially designed to run on the Raspberry Pi computer.

BRIEF CONTENTS

CONTENTS IN DETAIL

2
MAKING TEDIOUS ARITHMETIC FUN WITH LISTS AND LOOPS 19

3
GUESSING AND CHECKING WITH CONDITIONALS 37

PART 2: RIDING INTO MATH TERRITORY

4
TRANSFORMING AND STORING NUMBERS WITH ALGEBRA 53

5
TRANSFORMING SHAPES WITH GEOMETRY 77

6 CREATING OSCILLATIONS WITH TRIGONOMETRY 103

7 COMPLEX NUMBERS 127

8 USING MATRICES FOR COMPUTER GRAPHICS AND SYSTEMS OF EQUATIONS 145

PART 3: BLAZING YOUR OWN TRAIL

9
BUILDING OBJECTS WITH CLASSES 175

10
CREATING FRACTALS USING RECURSION 201

11
CELLULAR AUTOMATA 225

12
SOLVING PROBLEMS USING GENETIC ALGORITHMS 247

INDEX 273

ACKNOWLEDGMENTS

I'd like to thank Don "The Mathman" Cohen for showing me how fun and challenging learning real math can be; Seymour Papert for proving that coding belongs in math class; Mark Miller for giving me a chance to put my ideas into action; Hansel Lynn and Wayne Teng of theCoderSchool, who let me continue to have fun coding with students; and Ken Hawthorn for sharing my projects at his school. Thank you to my No Starch editors, Annie Choi, Liz Chadwick, and Meg Sneeringer, for all your help making this a much better book, and to Paddy Gaunt, whose input is visible all over this book. This book wouldn't exist without you all. Thank you to everybody who said no—you gave me the energy to keep going. Finally, thank you to Lucy for always believing in me.

INTRODUCTION

Which approach shown in Figure 1 would you prefer? On the left, you see an example of a traditional approach to teaching math, involving definitions, propositions, and proofs. This method requires a lot of reading and odd symbols. You'd never guess this had anything to do with geometric figures. In fact, this text explains how to find the *centroid*, or the center, of a triangle. But traditional approaches like this don't tell us *why* we should be interested in finding the center of a triangle in the first place.

434 SOLID GEOMETRY—BOOK VII

8. Given a regular hexagonal prism of height h and base edge e. $h = \frac{4}{3}e$. From this prism two congruent pyramids are to be removed, their bases being congruent to the bases of the prism. Find the height of either pyramid if the remaining solid is $\frac{3}{4}$ of the prism.

9. Centroids. The *center of gravity* of a material solid is commonly understood to be the point at which the weight of the solid acts. If a line, straight or curved, is regarded as a very thin homogeneous rod of uniform thickness, the center of gravity of the rod is called the **centroid** of the line. For example, the centroid of a straight line is its mid-point. Again, if a closed plane figure is regarded as a very thin homogeneous sheet (lamina) of uniform thickness, its center of gravity is called the **centroid** of the plane figure. In the same way, the **centroid** of a solid figure is defined as identical with the center of gravity of that figure, regarded as a homogeneous material solid. Thus the centroid of a spherical figure is its center.

By using the principle of the lever (Plane Geometry, Ex. 8, p. 234) it is easily shown that the center of gravity of a solid composed of two parts lies on the line-segment joining the centers of gravity of the parts and divides the segment in the inverse ratio of the weights of the parts. This result is readily extended to prove the following fundamental principle: *If a figure, plane or solid, is subdivided into parts whose centroids lie on a straight line, then the centroid of the entire figure lies on this line.*

Apply the preceding principle to a triangle by subdividing it into very narrow strips by drawing lines parallel to a side. Draw the median to this side. Then it is easy to infer that the centroid of the triangle lies on this median. Hence *the centroid of a triangle is the point of intersection of the medians.* (§§ 247–248.)

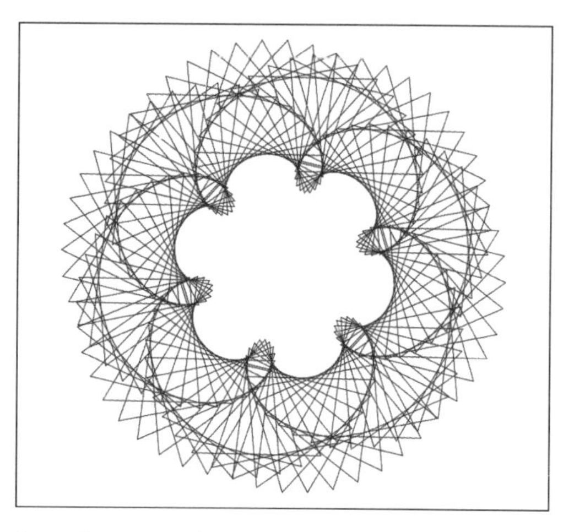

Figure 1: Two approaches to teaching about the centroid

Next to this text, you see a picture of a dynamic sketch with a hundred or so rotating triangles. It's a challenging programming project, and if you want it to rotate the right way (and look cool), you have to find the centroid of the triangle. In many situations, making cool graphics is nearly impossible without knowing the math behind geometry, for example. As you'll see in this book, knowing a little of the math behind triangles, like the centroid, will make it easy to create our artworks. A student who knows math and can create cool designs is more likely to delve into a little geometry and put up with a few square roots or a trig function or two. A student who doesn't see any outcome, and is only doing homework from a textbook, probably doesn't have much motivation to learn geometry.

In my eight years of experience as a math teacher and three years of experience as a computer science teacher, I've met many more math learners who prefer the visual approach to the academic one. In the process of creating something interesting, you come to understand that math is not just following steps to solve an equation. You see that exploring math with programming allows for many ways to solve interesting problems, with many unforeseen mistakes and opportunities for improvements along the way.

This is the difference between school math and real math.

THE PROBLEM WITH SCHOOL MATH

What do I mean by "school math" exactly? In the US in the 1860s, school math was preparation for a job as a clerk, adding columns of numbers by hand. Today, jobs are different, and the preparation for these jobs needs to change, too.

People learn best by doing. This hasn't been a daily practice in schools, though, which tend to favor passive learning. "Doing" in English and history classes might mean students write papers or give presentations, and science students perform experiments, but what do math students do? It used to be that all you could actively "do" in math class was solve equations, factor polynomials, and graph functions. But now that computers can do most of those calculations for us, these practices are no longer sufficient.

Simply learning how to automate solving, factoring, and graphing is not the final goal. Once a student has learned to automate a process, they can go further and deeper into a topic than was ever possible before.

Figure 2 shows a typical math problem you'd find in a textbook, asking students to define a function, "f(x)," and evaluate it for a ton of values.

Exercises 1 – 22 refer to the functions below. Find the indicated value of the function.

$$f(x) = \sqrt{x+3} - x + 1$$
$$g(t) = t^2 - 1$$
$$h(x) = x^2 + \frac{1}{x} + 2$$

1. $f(0)$
2. $f(1)$
3. $f(\sqrt{2})$
4. $f(\sqrt{2} - 1)$

Figure 2: A traditional approach to teaching functions

This same format goes on for 18 more questions! This kind of exercise is a trivial problem for a programming language like Python. We could simply define the function `f(x)` and then plug in the values by iterating over a list, like this:

```
import math

def f(x):
    return math.sqrt(x + 3) - x + 1

#list of values to plug in
for x in [0,1,math.sqrt(2),math.sqrt(2)-1]:
    print("f({:.3f}) = {:.3f}".format(x,f(x)))
```

The last line just makes the output pretty while rounding all the solutions to three decimal places, as shown here:

```
f(0.000) = 2.732
f(1.000) = 2.000
f(1.414) = 1.687
f(0.414) = 2.434
```

In programming languages like Python, JavaScript, Java, and so on, functions are a vitally important tool for transforming numbers and other objects—even other functions! Using Python, you can give a descriptive name to a function, so it's easier to understand what's going on. For example, you can name a function that calculates the area of a rectangle by calling it `calculateArea()`, like this:

```
def calculateArea(width,height):
```

A math textbook published in the 21st century, decades after Benoit Mandelbrot first generated his famous fractal on a computer when working for IBM, shows a picture of the Mandelbrot set and gushes over the discovery. The textbook describes the Mandelbrot set, which is shown in Figure 3, as "a fascinating mathematical object derived from the complex numbers. Its beautiful boundary illustrates chaotic behavior."

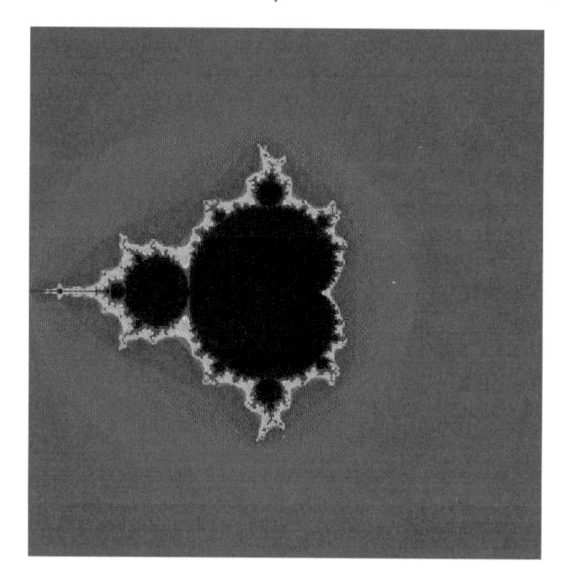

Figure 3: The Mandelbrot set

The textbook then takes the reader through a painstaking "exploration" to show how to transform a point in the complex plane. But the student is only shown how to do this on a calculator, which means only two points can be transformed (iterated seven times) in a reasonable amount of time. Two points.

In this book, you'll learn how to do this in Python, and you'll make the program transform hundreds of thousands of points automatically and even *create* the Mandelbrot set you see above!

ABOUT THIS BOOK

This book is about using programming tools to make math fun and relevant, while still being challenging. You'll make graphs to show all the possible outputs of a function. You'll make dynamic, interactive works of art. You'll even make an ecosystem with sheep that move around, eat grass, and multiply, and you'll create virtual organisms that try to find the shortest route through a bunch of cities while you watch!

You'll do this using Python and Processing in order to supercharge what you can do in math class. This book is not about skipping the math; it's about using the newest, coolest tools out there to get creative and learn real computer skills while discovering the connections between math, art, science, and technology. Processing will provide the graphics, shapes, motion, and colors, while Python does the calculating and follows your instructions behind the scenes.

For each of the projects in this book, you'll build the code up from scratch, starting from a blank file, and checking your progress at every step. Through making mistakes and debugging your own programs, you'll get a much deeper understanding of what each block of code does.

WHO SHOULD USE THIS BOOK

This book is for anyone who's learning math or who wants to use the most modern tools available to approach math topics like trigonometry and algebra. If you're learning Python, you can use this book to apply your growing programming skills to nontrivial projects like cellular automata, genetic algorithms, and computational art.

Teachers can use the projects in this book to challenge their students or to make math more approachable and relevant. What better way to teach matrices than to save a bunch of points to a matrix and use them to draw a 3D figure? When you know Python, you can do this and much more.

WHAT'S IN THIS BOOK?

This book begins with three chapters that cover basic Python concepts you'll build on to explore more complicated math. The next nine chapters explore math concepts and problems that you can visualize and solve using Python and Processing. You can try the exercises peppered throughout the book to apply what you learned and challenge yourself.

Chapter 1: Drawing Polygons with Turtles teaches basic programming concepts like loops, variables, and functions using Python's built-in `turtle` module.

Chapter 2: Making Tedious Arithmetic Fun with Lists and Loops goes deeper into programming concepts like lists and Booleans.

Chapter 3: Guessing and Checking with Conditionals applies your growing Python skills to problems like factoring numbers and making an interactive number-guessing game.

Chapter 4: Transforming and Storing Numbers with Algebra ramps up from solving simple equations to solving cubic equations numerically and by graphing.

Chapter 5: Transforming Shapes with Geometry shows you how to create shapes and then multiply, rotate, and spread them all over the screen.

Chapter 6: Creating Oscillations with Trigonometry goes beyond right triangles and lets you create oscillating shapes and waves.

Chapter 7: Complex Numbers teaches you how to use complex numbers to move points around the screen, creating designs like the Mandelbrot set.

Chapter 8: Using Matrices for Computer Graphics and Systems of Equations takes you into the third dimension, where you'll translate and rotate 3D shapes and solve huge systems of equations with one program.

Chapter 9: Building Objects with Classes covers how to create one object, or as many as your computer can handle, with roaming sheep and delicious grass locked in a battle for survival.

Chapter 10: Creating Fractals Using Recursion shows how recursion can be used as a whole new way to measure distances and create wildly unexpected designs.

Chapter 11: Cellular Automata teaches you how to generate and program cellular automata to behave according to rules you make.

Chapter 12: Solving Problems Using Genetic Algorithms shows you how to harness the theory of natural selection to solve problems we couldn't solve in a million years otherwise!

DOWNLOADING AND INSTALLING PYTHON

The easiest way to get started is to use the Python 3 software distribution, which is available for free at *https://www.python.org/*. Python has become one of the most popular programming languages in the world. It's used to create websites like Google, YouTube, and Instagram, and researchers at universities all over the world use it to crunch numbers in various fields, from astronomy to zoology. The latest version released to date is Python 3.7. Go to *https://www.python.org/downloads/* and choose the latest version of Python 3, as shown in Figure 4.

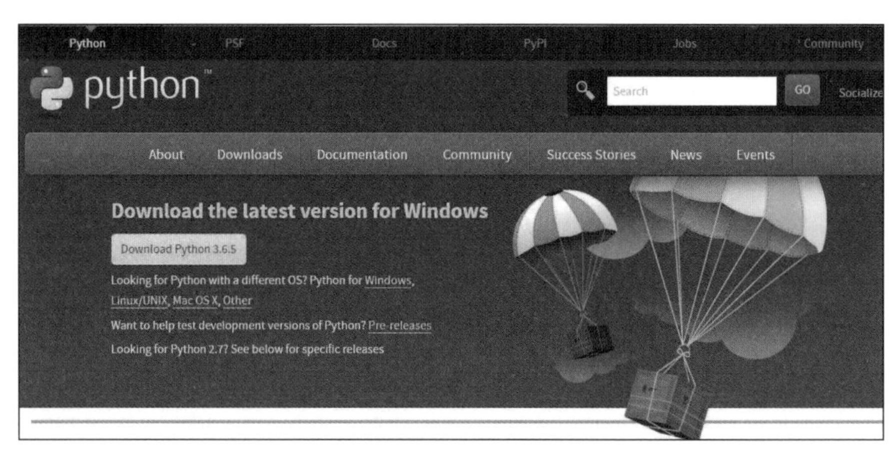

Figure 4: The official website of the Python Software Foundation

You can choose the version for your operating system. The site detected that I was using Windows. Click the file when the download is complete, as shown in Figure 5.

Figure 5: Click the downloaded file to start the install

Follow the directions, and always choose the default options. It might take a few minutes to install. After that, search your system for "IDLE." That's the Python IDE, or *integrated development environment*, which is what you'll need to write Python code. Why "IDLE"? The Python programming language was named after the Monty Python comedy troupe, and one of the members is Eric Idle.

STARTING IDLE

Find IDLE on your system and open it.

Figure 6: Opening IDLE on Windows

A screen called a "shell" will appear. You can use this for the interactive coding environment, but you'll want to save your code. Click **File ▸ New File** or press ALT-N, and a file will appear (see Figure 7).

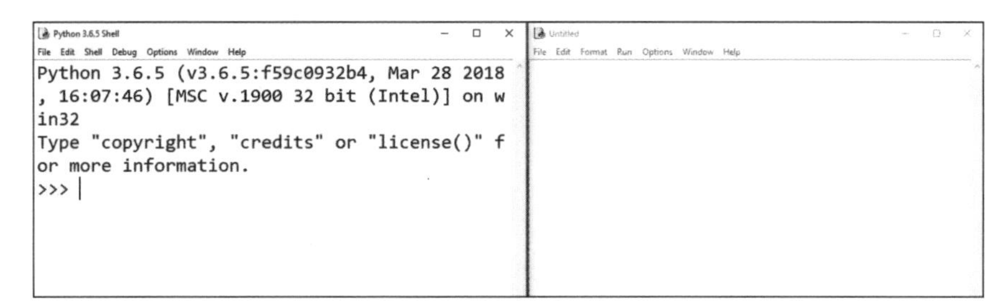

Figure 7: Python's interactive shell (left) and a new module (file) window, ready for code!

This is where you'll write your Python code. We will also use Processing, so let's go over how to download and install Processing next.

INSTALLING PROCESSING

There's a lot you can do with Python, and we'll use IDLE a lot. But when we want to do some heavy-duty graphics, we're going to use Processing. Processing is a professional-level graphics library used by coders and artists to make dynamic, interactive artwork and graphics.

Go to *https://processing.org/download/* and choose your operating system, as shown in Figure 8.

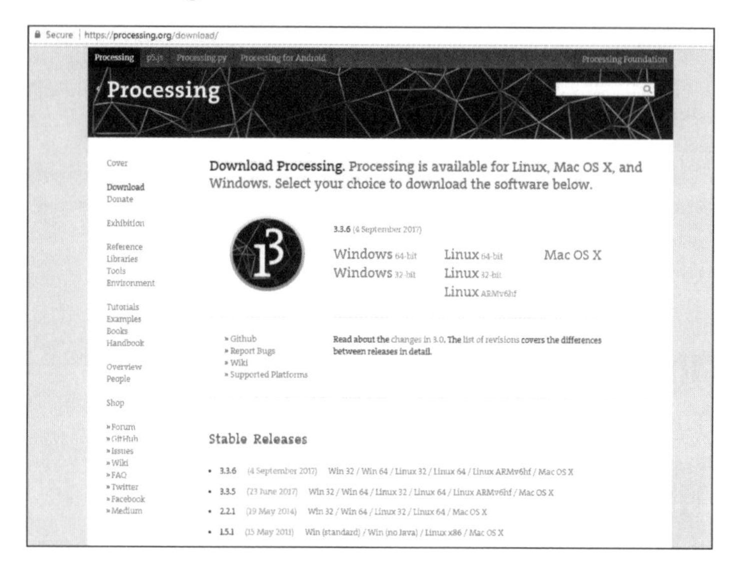

Figure 8: The Processing website

Download the installer for your operating system by clicking it and following the instructions. Double-click the icon to start Processing. This defaults to Java mode. Click **Java** to open the drop-down menu, as shown in Figure 9, and then click **Add Mode**.

Select **Python Mode ▸ Install**. It should take a minute or two, but after this you'll be able to code in Python with Processing.

Now that you've set up Python and Processing, you're ready to start exploring math!

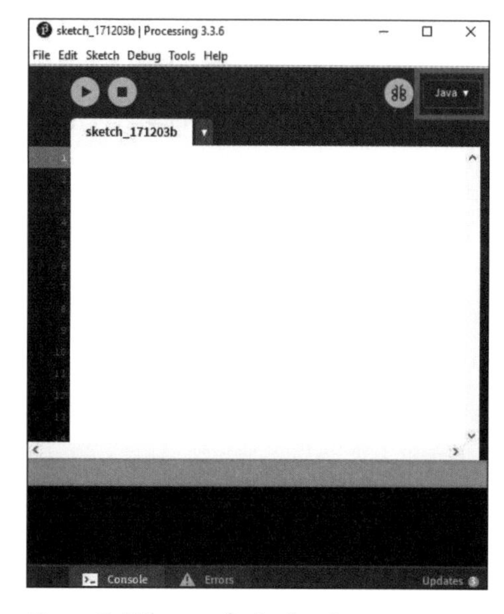

Figure 9: Where to find other Processing modes, like the Python mode we'll be using

PART I

HITCHIN' UP YOUR PYTHON WAGON

DRAWING POLYGONS WITH THE TURTLE MODULE

Centuries ago a Westerner heard a Hindu say the Earth rested on the back of a turtle. When asked what the turtle was standing on, the Hindu explained, "It's turtles all the way down."

Before you can start using math to build all the cool things you see in this book, you'll need to learn how to give instructions to your computer using a programming language called Python. In this chapter you'll get familiar with some basic programming concepts like loops, variables, and functions by using Python's built-in turtle tool to draw different shapes. As you'll see, the turtle module is a fun way to learn about Python's basic features and get a taste of what you'll be able to create with programming.

PYTHON'S TURTLE MODULE

The Python turtle tool is based on the original "turtle" agent from the Logo programming language, which was invented in the 1960s to make computer programming more accessible to everyone. Logo's graphical environment made interacting with the computer visual and engaging. (Check out Seymour Papert's brilliant book *Mindstorms* for more great ideas for learning math using Logo's virtual turtles.) The creators of the Python programming language liked the Logo turtles so much that they wrote a module called *turtle* in Python to copy the Logo turtle functionality.

Python's turtle module lets you control a small image shaped like a turtle, just like a video game character. You need to give precise instructions to direct the turtle around the screen. Because the turtle leaves a trail wherever it goes, we can use it to write a program that draws different shapes.

Let's begin by importing the turtle module!

IMPORTING THE TURTLE MODULE

Open a new Python file in IDLE and save it as *myturtle.py* in the Python folder. You should see a blank page. To use turtles in Python, you have to import the functions from the turtle module first.

A *function* is a set of reusable code for performing a specific action in a program. There are many built-in functions you can use in Python, but you can also write your own functions (you'll learn how to write your own functions later in this chapter).

A *module* in Python is a file that contains predefined functions and statements that you can use in another program. For example, the turtle module contains a lot of useful code that was automatically downloaded when you installed Python.

Although functions can be imported from a module in many ways, we'll use a simple one here. In the *myturtle.py* file you just created, enter the following at the top:

```
from turtle import *
```

The `from` command indicates that we're importing something from outside our file. We then give the name of the module we want to import from, which is `turtle` in this case. We use the `import` keyword to get the useful code we want *from* the turtle module. We use the asterisk (*) here as a *wildcard command* that means "import everything from that module." Make sure to put a space between `import` and the asterisk.

Save the file and make sure it's in the Python folder; otherwise, the program will throw an error.

WARNING *Do not save the file as* turtle.py. *This filename already exists and will cause a conflict with the import from the turtle module! Anything else will work:* myturtle.py, turtle2.py, mondayturtle.py, *and so on.*

MOVING YOUR TURTLE

Now that you've imported the turtle module, you're ready to enter instructions to move the turtle. We'll use the `forward()` function (abbreviated as `fd`) to move the turtle forward a certain number of steps while leaving a trail behind it. Note that `forward()` is one of the functions we just imported from the turtle module. Enter the following to make the turtle go forward:

```
forward(100)
```

Here, we use the `forward()` function with the number 100 inside parentheses to indicate how many steps the turtle should move. In this case, 100 is the *argument* we pass to the `forward()` function. All functions take one or more arguments. Feel free to pass other numbers to this function. When you press F5 to run the program, a new window should open with an arrow in the center, as shown in Figure 1-1.

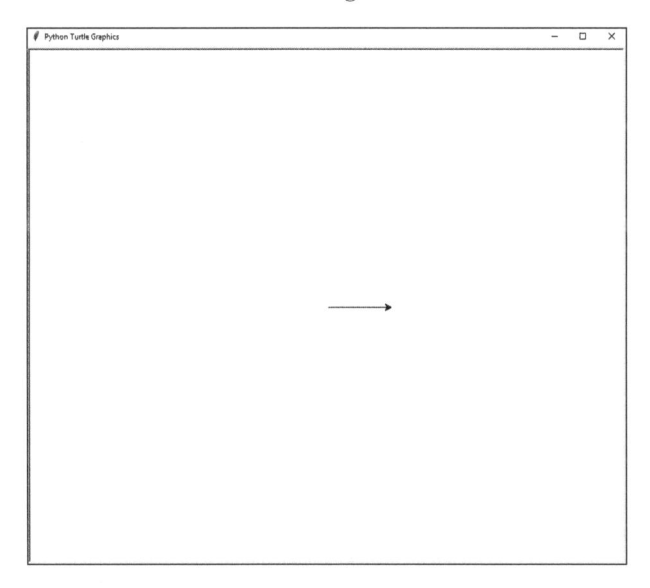

Figure 1-1: Running your first line of code!

As you can see, the turtle started in the middle of the screen and walked forward 100 steps (it's actually 100 pixels). Notice that the default shape is an arrow, not a turtle, and the default direction the arrow is facing is to the right. To change the arrow into a turtle, update your code so that it looks like this:

myturtle.py

```
from turtle import *
forward(100)
shape('turtle')
```

As you can probably tell, `shape()` is another function defined in the turtle module. It lets you change the shape of the default arrow into other shapes, like a circle, a square, or an arrow. Here, the `shape()` function takes the string value 'turtle' as its argument, not a number. (You'll learn more about strings and different data types in the next chapter.) Save and run the *myturtle.py* file again. You should see something like Figure 1-2.

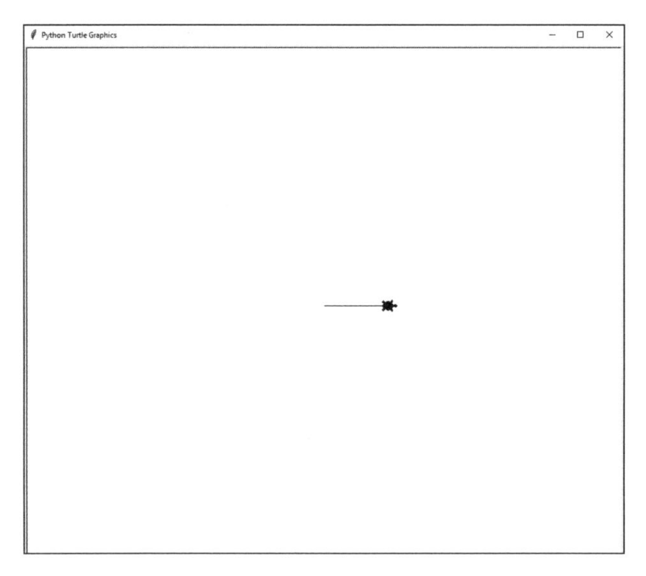

Figure 1-2: Changing the arrow into a turtle!

Now your arrow should look like a tiny turtle!

CHANGING DIRECTIONS

The turtle can go only in the direction it's facing. To change the turtle's direction, you must first make the turtle turn a specified number of degrees using the `right()` or `left()` function and then go forward. Update your *myturtle.py* program by adding the last two lines of code shown next:

myturtle.py

```
from turtle import *
forward(100)
shape('turtle')
right(45)
forward(150)
```

Here, we'll use the `right()` function (or `rt()` for short) to make the turtle turn right 45 degrees before moving forward by 150 steps. When you run this code, the output should look like Figure 1-3.

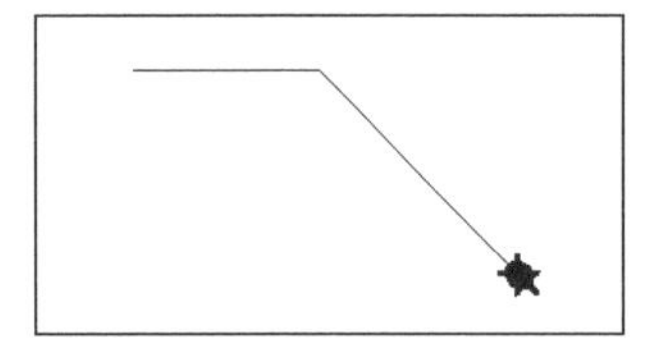

Figure 1-3: Changing turtle's direction

As you can see, the turtle started in the middle of the screen, went forward 100 steps, turned right 45 degrees, and then went forward another 150 steps. Notice that Python runs each line of code in order, from top to bottom.

> **EXERCISE 1-1: SQUARE DANCE**
>
> Return to the *myturtle.py* program. Your first challenge is to modify the code in the program using only the `forward` and `right` functions so that the turtle draws a square.

REPEATING CODE WITH LOOPS

Every programming language has a way to automatically repeat commands a given number of times. This is useful because it saves you from having to type out the same code over and over and cluttering your program. It also helps you avoid typos that can prevent your program from running properly.

USING THE FOR LOOP

In Python we use the `for` loop to repeat code. We also use the `range` keyword to specify the number of times we go through the loop. Open a new program file in IDLE, save it as *for_loop.py*, and then enter the following:

for_loop.py

```
for i in range(2):
    print('hello')
```

Here, the `range()` function creates `i`, or an *iterator*, for each for loop. The iterator is a value that increases each time it's used. The number 2 in parentheses is the argument we pass to the function to control its behavior. This is similar to the way we passed different values to the `forward()` and `right()` functions in previous sections.

In this case, `range(2)` creates a sequence of two numbers, 0 and 1. For each of these two numbers, the `for` command performs the action specified after the colon, which is to print the word *hello*.

Be sure to indent all the lines of the code you want to repeat by pressing TAB (one tab is four spaces). Indentation tells Python which lines are inside the loop so `for` knows exactly what code to repeat. And don't forget the colon at the end; it tells the computer what's coming up after it is in the loop. When you run the program, you should see the following printed in the shell:

```
hello
hello
```

As you can see, the program prints `hello` twice because `range(2)` creates a sequence containing two numbers, 0 and 1. This means that the `for` command loops over the two items in the sequence, printing "hello" each time. Let's update the number in the parentheses, like this:

for_loop.py

```
for i in range(10):
    print('hello')
```

When you run this program, you should get `hello` ten times, like this:

```
hello
hello
hello
hello
hello
hello
hello
hello
hello
hello
```

Let's try another example since you'll be writing a lot of `for` loops in this book:

for_loop.py

```
for i in range(10):
    print(i)
```

Because counting begins at 0 rather than 1 in Python, `for i in range(10)` gives us the numbers 0 through 9. This sample code is saying "for each value in the range 0 to 9, display the current number." The `for` loop then repeats the code until it runs out of numbers in the range. When you run this code, you should get something like this:

```
0
1
2
3
4
5
6
7
8
9
```

In the future you'll have to remember that `i` starts at 0 and ends before the last number in a loop using `range`, but for now, if you want something repeated four times, you can use this:

```
for i in range(4):
```

It's as simple as that! Let's see how we can put this to use.

USING A FOR LOOP TO DRAW A SQUARE

In Exercise 1-1 your challenge was to make a square using only the `forward()` and `right()` functions. To do this, you had to repeat `forward(100)` and `right(90)` four times. But this required entering the same code multiple times, which is time-consuming and can lead to mistakes.

Let's use a `for` loop to avoid repeating the same code. Here's the *myturtle .py* program, which uses a `for` loop instead of repeating the `forward()` and `right()` functions four times:

myturtle.py

```
from turtle import *
shape('turtle')
for i in range(4):
    forward(100)
    right(90)
```

Note that `shape('turtle')` should come right after you import the turtle module and before you start drawing. The two lines of code inside this `for` loop tell the turtle to go forward 100 steps and then turn 90 degrees to the right. (You might have to face the same way as the turtle to know which way "right" is!) Because a square has four sides, we use `range(4)` to repeat these two lines of code four times. Run the program, and you should see something like Figure 1-4.

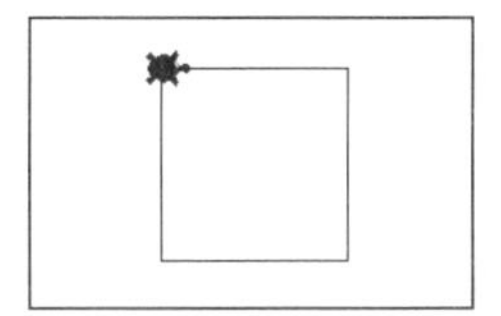

Figure 1-4: A square made with a `for` loop

You should see that the turtle moves forward and turns to the right a total of four times, finally returning to its original position. You successfully drew a square using a `for` loop!

CREATING SHORTCUTS WITH FUNCTIONS

Now that we've written code to draw a square, we can save all that code to a magic keyword that we can call any time we want to use that square code again. Every programming language has a way to do this, and in Python it's called a *function*, which is the most important feature of computer programming. Functions make code compact and easier to maintain, and dividing a problem up into functions often allows you to see the best way of solving it. Earlier you used some built-in functions that come with the turtle module. In this section you learn how to define your own function.

To define a function you start by giving it a name. This name can be anything you want, as long as it's not already a Python keyword, like `list`, `range`, and so on. When you're naming functions, it's better to be descriptive so you can remember what they're for when you use them again. Let's call our function `square()` because we'll be using it to make a square:

myturtle.py

```
def square():
    for i in range(4):
        forward(100)
        right(90)
```

The `def` command tells Python we're defining a function, and the word we list afterward will become the function name; here, it's `square()`. Don't forget the parentheses after `square`! They're a sign in Python that you're dealing with a function. Later we'll put values inside them, but even without any values inside, the parentheses need to be included to let Python know you are defining a function. Also, don't forget the colon at the end of the function definition. Note that we indent all the code inside the function to let Python know which code goes inside it.

If you run this program now, nothing will happen. You've defined a function, but you didn't tell the program to run it yet. To do this, you need to *call* the function at the end of the *myturtle.py* file after the function definition. Enter the code shown in Listing 1-1.

myturtle.py

```
from turtle import *
shape('turtle')
def square():
    for i in range(4):
        forward(100)
        right(90)
square()
```

Listing 1-1: The `square()` function is called at the end of the file.

When you call `square()` at the end like this, the program should run properly. Now you can use the `square()` function at any point later in the program to quickly draw another square.

You can also use this function in a loop to build something more complicated. For example, to draw a square, turn right a little, make another square, turn right a little, and repeat those steps multiple times, putting the function inside a loop makes sense.

The next exercise shows an interesting-looking shape that's made of squares! It might take your turtle a while to create this shape, so you can speed it up by adding the `speed()` function to *myturtle.py* after `shape('turtle')`. Using `speed(0)` makes the turtle move the fastest, whereas `speed(1)` is the slowest. Try different speeds, like `speed(5)` and `speed(10)`, if you want.

EXERCISE 1-2: A CIRCLE OF SQUARES

Write and run a function that draws 60 squares, turning right 5 degrees after each square. Use a loop! Your result should end up looking like this:

USING VARIABLES TO DRAW SHAPES

So far all our squares are the same size. To make squares of different sizes, we'll need to vary the distance the turtle walks forward for each side. Instead of changing the definition for the `square()` function every time we want a different size, we can use a *variable*, which in Python is a word that represents a value you can change. This is similar to the way *x* in algebra can represent a value that can change in an equation.

In math class, variables are single letters, but in programming you can give a variable any name you want! Like with functions, I suggest naming variables something descriptive to make reading and understanding your code easier.

USING VARIABLES IN FUNCTIONS

When you define a function, you can use variables as the function's parameters inside the parentheses. For example, you can change your `square()` function definition in the *myturtle.py* program to the following to create squares of any size rather than a fixed size:

myturtle.py

```
def square(sidelength):
    for i in range(4):
        forward(sidelength)
        right(90)
```

Here, we use `sidelength` to define the `square()` function. Now when you call this function, you have to place a value, which we call an *argument*, inside the parentheses, and whatever number is inside the parentheses will be used in place of `sidelength`. For example, calling `square(50)` and `square(80)` would look like Figure 1-5.

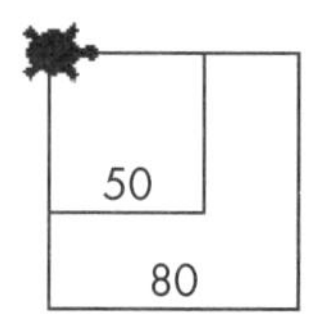

Figure 1-5: A square of size 50 and a square of size 80

When you use a variable to define a function, you can simply call the square() function by entering different numbers without having to update the function definition each time.

VARIABLE ERRORS

At the moment, if we forget to put a value in the parentheses for the function, we'll get this error:

```
Traceback (most recent call last):
  File "C:/Something/Something/my_turtle.py", line 12, in <module>
    square()
TypeError: square() missing 1 required positional argument: 'sidelength'
```

This error tells us that we're missing a value for sidelength, so Python doesn't know how big to make the square. To avoid this, we can give a default value for the length in the first line of the function definition, like this:

```
def square(sidelength=100):
```

Here, we place a default value of 100 in sidelength. Now if we put a value in the parentheses after square, it'll make a square of that length, but if we leave the parentheses empty, it'll default to a square of sidelength 100 and won't give us an error. The updated code should produce the drawing shown in Figure 1-6:

```
square(50)
square(30)
square()
```

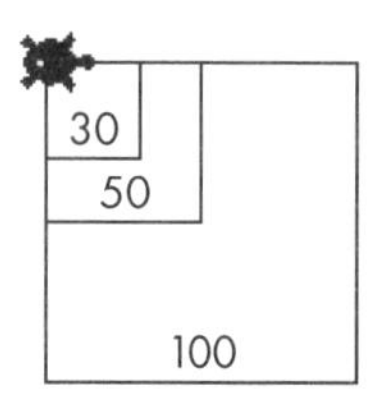

Figure 1-6: A default square of size 100, a square of size 50, and a square of size 30

Setting a default value like this makes it easier to use our function without having to worry about getting errors if we do something wrong. In programming this is called making the program more *robust*.

> **EXERCISE 1-3: TRI AND TRI AGAIN**
>
> Write a `triangle()` function that will draw a triangle of a given "side length."

EQUILATERAL TRIANGLES

A *polygon* is a many-sided figure. An *equilateral triangle* is a special type of polygon that has three equal sides. Figure 1-7 shows what it looks like.

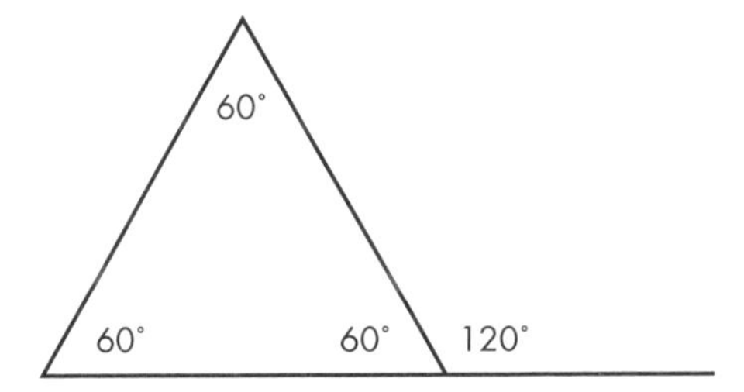

Figure 1-7: The angles in an equilateral triangle, including one external angle

An equilateral triangle has three equal internal angles of 60 degrees. Here's a rule you might remember from geometry class: all three angles of an equilateral triangle add up to 180 degrees. In fact, this is true for all triangles, not just equilateral triangles.

WRITING THE TRIANGLE() FUNCTION

Let's use what you've learned so far to write a function that makes the turtle walk in a triangular path. Because each angle in an equilateral triangle is 60 degrees, you can update the `right()` movement in your `square()` function to `60`, like this:

myturtle.py

```
def triangle(sidelength=100):
    for i in range(3):
        forward(sidelength)
        right(60)

triangle()
```

But when you save and run this program, you won't get a triangle. Instead, you'll see something like Figure 1-8.

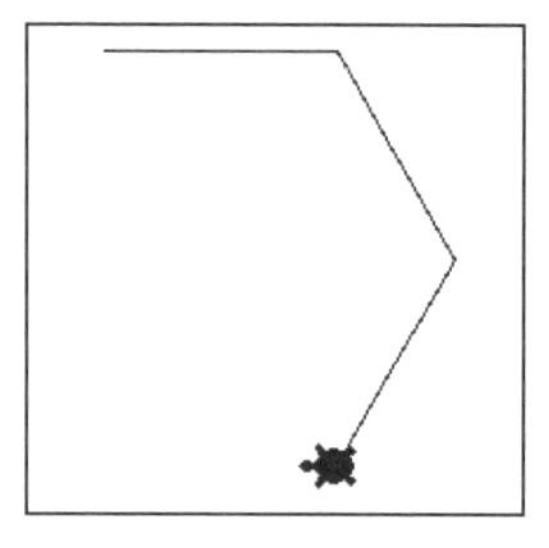

Figure 1-8: A first attempt at drawing a triangle

That looks like we're starting to draw a hexagon (a six-sided polygon), not a triangle. We get a hexagon instead of a triangle because we entered 60 degrees, which is the *internal* angle of an equilateral triangle. We need to enter the *external* angle to the `right()` function instead, because the turtle *turns* the external angle, not the internal angle. This wasn't a problem with the square because it just so happens the internal angle of a square and the external angle are the same: 90 degrees.

To find the external angle for a triangle, simply subtract the internal angle from 180. This means the external angle of an equilateral triangle is 120 degrees. Update 60 in the code to 120, and you should get a triangle.

EXERCISE 1-4: POLYGON FUNCTIONS

Write a function called `polygon` that takes an integer as an argument and makes the turtle draw a polygon with that integer's number of sides.

MAKING VARIABLES VARY

There's more we can do with variables: we can automatically increase the variable by a certain amount so that each time we run the function, the square is bigger than the last. For example, using a `length` variable, we can make a square, then increase the `length` variable a little before making the next square by incrementing the variable like this:

```
length = length + 5
```

As a math guy, this line of code didn't make sense to me when I first saw it! How can "length equal length + 5"? It's not possible! But code isn't an equation, and an equal sign (`=`) in this case doesn't mean "this side equals that side." *The equal sign in programming means we're assigning a value.*

Take the following example. Open the Python shell and enter the following code:

```
>>> radius = 10
```

This means we're creating a variable called radius (if there isn't one already) and assigning it the value 10. You can always assign a different value to it later, like this:

```
radius = 20
```

Press ENTER and your code will be executed. This means the value 20 will be assigned to the radius variable. To check whether a variable is equal to something, use double equal signs (==). For example, to check whether the value of the radius variable is 20, you can enter this into the shell:

```
>>> radius == 20
```

Press ENTER and it should print the following:

```
True
```

Now the value of the radius variable is 20. It's often useful to increment variables rather than assign them number values manually. You can use a variable called count to count how many times something happens in a program. It should start at 0 and go up by one after every occurrence. To make a variable go up by one in value, you add 1 to its value and then assign the new value to the variable, like this:

```
count = count + 1
```

You can also write this as follows to make the code more compact:

```
count += 1
```

This means "add 1 to my count variable." You can use addition, subtraction, multiplication, and division in this notation. Let's see it in action by running this code in the Python shell. We'll assign x the value 12 and y the value 3, and then make x go up by y:

```
>>> x = 12
>>> y = 3
>>> x += y
>>> x
15
>>> y
3
```

Notice y didn't change. We can increment x using addition, subtraction, multiplication, and division with similar notation:

```
>>> x += 2
>>> x
17
```

Now we'll set x to one less than its current value:

```
>>> x -= 1
>>> x
16
```

We know that x is 16. Now let's set x to two times its current value:

```
>>> x *= 2
>>> x
32
```

Finally, we can set x to a quarter of its value by dividing it by 4:

```
>>> x /= 4
>>> x
8.0
```

Now you know how to increment a variable using arithmetic operators followed by an equal sign. In sum, `x += 3` will make `x` go up by 3, whereas `x -= 1` will make it go down by 1, and so on.

You can use the following line of code to make the length increment by 5 every loop, which will come in handy in the next exercises:

```
length += 5
```

With this notation, every time the `length` variable is used, 5 is added to the value and saved into the variable.

EXERCISE 1-5: TURTLE SPIRAL

Make a function to draw 60 squares, turning 5 degrees after each square and making each successive square bigger. Start at a `length` of 5 and increment 5 units every square. It should look like this:

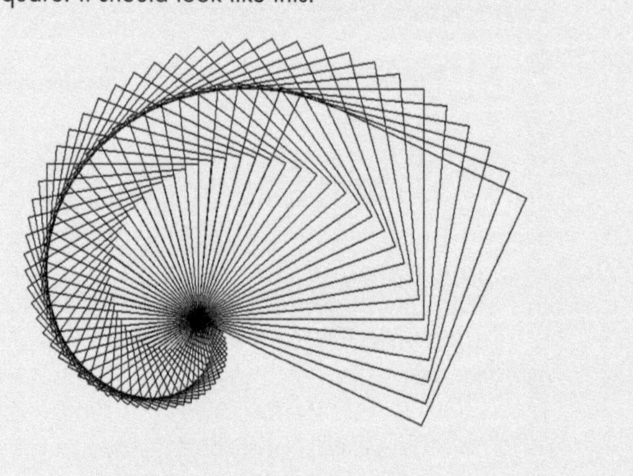

SUMMARY

In this chapter you learned how to use Python's turtle module and its built-in functions like `forward()` and `right()` to draw different shapes. You also saw that the turtle can perform many more functions than those we covered here. There are dozens more that I encourage you to experiment with before moving on to the next chapter. If you do a web search for "python turtle," the first result will probably be the turtle module documentation on the official Python website (*https://ptyhon.org/*) website. You'll find all the turtle methods on that page, some of which is shown in Figure 1-9.

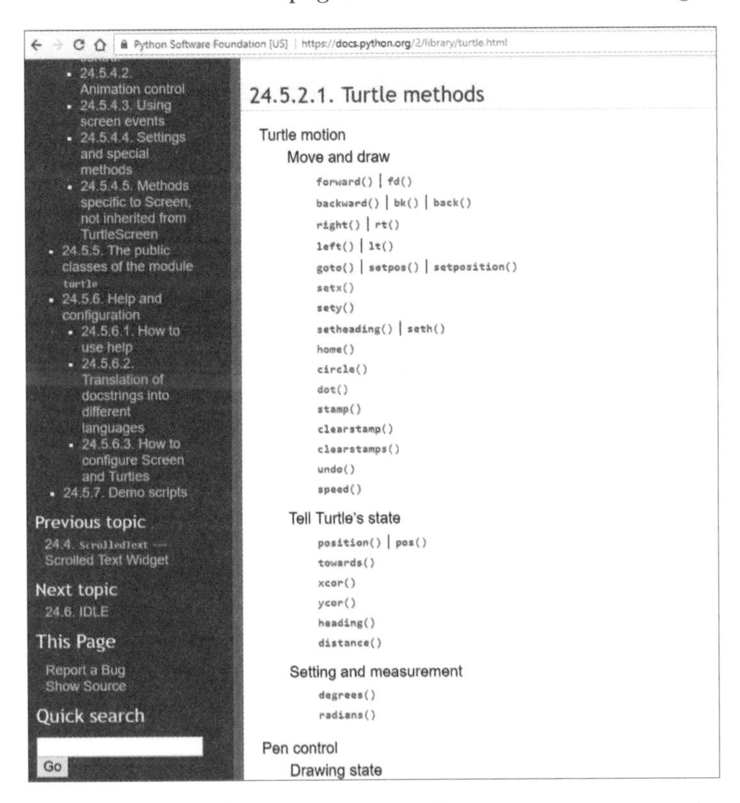

Figure 1-9: You can find many more turtle functions and methods on the Python website!

You learned how to define your own functions, thus saving valuable code that can be reused at any time. You also learned how to run code multiple times using `for` loops without having to rewrite the code. Knowing how to save time and avoid mistakes using functions and loops will be useful when you build more complicated math tools later on.

In the next chapter we'll build on the basic arithmetic operators you used to increment variables. You'll learn more about the basic operators and data types in Python and how to use them to build simple computation tools. We'll also explore how to store items in lists and use indices to access list items.

EXERCISE 1-6: A STAR IS BORN

First, write a "star" function that will draw a five-pointed star, like this:

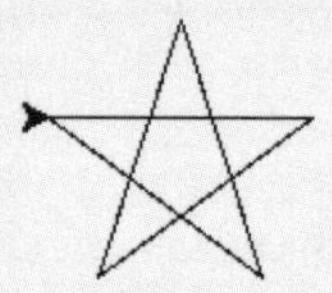

Next, write a function called `starSpiral()` that will draw a spiral of stars, like this:

MAKING TEDIOUS ARITHMETIC FUN WITH LISTS AND LOOPS

"You mean I have to go again tomorrow?"
—Aidan Farrell after the first day of school

Most people think of doing arithmetic when they think of math: adding, subtracting, multiplying, and dividing. Although doing arithmetic is pretty easy using calculators and computers, it can still involve a lot of repetitive tasks. For example, to add 20 different numbers using a calculator, you have to enter the + operator 19 times!

In this chapter you learn how to automate some of the tedious parts of arithmetic using Python. First, you learn about math operators and the different data types you can use in Python. Then you learn how to store and calculate values using variables. You also learn to use lists and loops to repeat code. Finally, you combine these programming concepts to write functions that automatically perform complicated calculations for you. You'll see that Python can be a much more powerful calculator than any calculator you can buy—and best of all, it's free!

BASIC OPERATORS

Doing arithmetic in the interactive Python shell is easy: you just enter the expression and press ENTER when you want to do the calculation. Table 2-1 shows some of the most common mathematical operators.

Table 2-1: Common Mathematical Operators in Python

Operator	Syntax
Addition	+
Subtraction	–
Multiplication	*
Division	/
Exponent	**

Open your Python shell and try out some basic arithmetic with the example in Listing 2-1.

```
>>> 23 + 56  #Addition
79
>>> 45 * 89  #Multiplication is with an asterisk
4005
>>> 46 / 13  #Division is with a forward slash
3.5384615384615383
>>> 2 ** 4   #2 to the 4th power
16
```

Listing 2-1: Trying out some basic math operators

The answer should appear as the output. You can use spaces to make the code more readable (`6 + 5`) or not (`6+5`), but it won't make any difference to Python when you're doing arithmetic.

Keep in mind that division in Python 2 is a little tricky. For example, Python 2 will take `46/13` and think you're interested only in integers, thus giving you a whole number (3) for the answer instead of returning a decimal value, like in Listing 2-1. Because you downloaded Python 3, you shouldn't have that problem. But the graphics package we'll see later uses Python 2, so we'll have to make sure we ask for decimals when we divide.

OPERATING ON VARIABLES

You can also use operators on variables. In Chapter 1 you learned to use variables when defining a function. Like variables in algebra, variables in programming allow long, complicated calculations to be broken into several stages by storing results that can be used again later. Listing 2-2 shows how you can use variables to store numbers and operate on them, no matter what their value is.

```
>>> x = 5
>>> x = x + 2
>>> length = 12
>>> x + length
19
```

Listing 2-2: Storing results in variables

Here, we assign the value 5 to the x variable, then increment it by 2, so x becomes 7. We then assign the value 12 to the variable length. When we add x and length, we're adding 7 + 12, so the result is 19.

USING OPERATORS TO WRITE THE AVERAGE() FUNCTION

Let's practice using operators to find the mean of a series of numbers. As you may know from math class, to find the mean you add all the numbers together and divide them by how many numbers there are in the series. For example, if your numbers are 10 and 20, you add 10 and 20 and divide the sum by 2, as shown here:

$$(10 + 20) / 2 = 15$$

If your numbers are 9, 15, and 23, you add them together and divide the sum by 3:

$$(9 + 15 + 23) / 3 = 47 / 3 = 15.67$$

This can be tedious to do by hand but simple to do with code. Let's start a Python file called *arithmetic.py* and write a function to find the average of two numbers. You should be able to run the function and give it two numbers as arguments, without any operators, and have it print the average, like this:

```
>>> average(10,20)
15.0
```

Let's give it a try.

MIND THE ORDER OF OPERATIONS!

Our average() function transforms two numbers, a and b, into half their sum and then returns that value using the return keyword. Here's the code for our function:

arithmetic.py

```
def average(a,b):
    return a + b / 2
```

We define a function called average(), which requires two numbers, a and b, as inputs. We write that the function should return the sum of the two numbers divided by 2. However, when we test the function in the shell, we get the wrong output:

```
>>> average(10,20)
20.0
```

That's because we didn't take the *order of operations* into account when writing our function. As you probably remember from math class, multiplication and division take precedence over addition and subtraction, so in this case division is performed first. This function is dividing b by 2 and *then* adding a. So how do we fix this?

USING PARENTHESES WITH OPERATORS

We need to use parentheses to tell Python to add the two numbers first, before dividing:

arithmetic.py

```
def average(a,b):
    return (a + b) / 2
```

Now the function should add a and b before dividing by 2. Here's what happens when we run the function in the shell:

```
>>> average(10,20)
15.0
```

If you perform this same calculation by hand, you can see the output is correct! Try the average() function using different numbers.

DATA TYPES IN PYTHON

Before we continue doing arithmetic on numbers, let's explore some basic Python data types. Different data types have different capabilities, and you can't always perform the same operations on all of them, so it's important to know how each data type works.

INTEGERS AND FLOATS

Two Python data types you commonly perform operations on are integers and floats. *Integers* are whole numbers. *Floats* are numbers containing decimals. You can change integers to floats, and vice versa, by using the float() and int() functions, respectively, like so:

```
>>> x = 3
>>> x
3
>>> y = float(x)
>>> y
3.0
```

```
>>> z = int(y)
>>> z
3
```

In this example we use `x = 3` to assign the value 3 to the variable `x`. We then convert `x` into a float using `float(x)` and assign the result (`3.0`) to the variable `y`. Finally, we convert `y` into an integer and assign the result (`3`) to the variable `z`. This shows how you can easily switch between floats and ints.

STRINGS

Strings are ordered alphanumeric characters, which can be a series of letters, like words, or numbers. You define a string by enclosing the characters in single (`''`) or double quotes (`""`), like so:

```
>>> a = "hello"
>>> a + a
'hellohello'
>>> 4*a
'hellohellohellohello'
```

Here, we store the string `"hello"` in variable `a`. When we add variable `a` to itself, we get a new string, `'hellohello'`, which is a combination of two `hello`s. Keep in mind that you can't add strings and number data types (integers and floats) together, though. If you try adding the integer `2` and the string `"hello"`, you'll get this error message:

```
>>> b = 2
>>> b
2
>>> d = "hello"
>>> b + d
Traceback (most recent call last):
  File "<pyshell#34>", line 1, in <module>
    b + d
TypeError: unsupported operand type(s) for +: 'int' and 'str'
```

However, if a number is a string (or enclosed in quotes), you can add it to another string, like this:

```
>>> b = '123'
>>> c = '4'
>>> b + c
'1234'
>>> 'hello' + ' 123'
'hello 123'
```

In this example both `'123'` and `'4'` are strings made up of numbers, not number data types. So when you add the two together you get a longer string (`'1234'`) that is a combination of the two strings. You can do the same with the strings 'hello' and ' 123', even though one is made of letters and the other is made of numbers. Joining strings to create a new string is called *concatenation*.

You can also multiply a string by an integer to repeat the string, like this:

```
>>> name = "Marcia"
>>> 3 * name
'MarciaMarciaMarcia'
```

But you can't subtract, multiply, or divide a string by another string. Enter the following in the shell to see what happens:

```
>>> noun = 'dog'
>>> verb = 'bark'
>>> noun * verb
Traceback (most recent call last):
  File "<pyshell#6>", line 1, in <module>
    noun * verb
TypeError: can't multiply sequence by non-int of type 'str'
```

As you can see, when you try to multiply 'dog' and 'bark', you get an error telling you that you can't multiply two string data types.

BOOLEANS

Booleans are true/false values, which means they can be only one or the other and nothing in between. Boolean values have to be capitalized in Python and are often used to compare the values of two things. To compare values you can use the greater-than (>) and less-than (<) symbols, like so:

```
>>> 3 > 2
True
```

Because 3 is greater than 2, this expression returns True. But checking whether two values are equal requires two equal signs (==), because one equal sign simply assigns a value to a variable. Here's an example of how this works:

```
>>> b = 5
>>> b == 5
True
>>> b == 6
False
```

First we assign the value 5 to variable b using one equal sign. Then we use two equal signs to check whether b is equal to 5, which returns True.

CHECKING DATA TYPES

You can always check which data type you're dealing with by using the type() function with a variable. Python conveniently tells you what data type the value in the variable is. For example, let's assign a Boolean value to a variable, like this:

```
>>> a = True
>>> type(a)
<class 'bool'>
```

When you pass variable a into the type() function, Python tells you that the value in a is a Boolean.

Try checking the data type of an integer:

```
>>> b = 2
>>> type(b)
<class 'int'>
```

The following checks whether 0.5 is a float:

```
>>> c = 0.5
>>> type(c)
<class 'float'>
```

This example confirms that alphanumeric symbols inside quotes are a string:

```
>>> name = "Steve"
>>> type(name)
<class 'str'>
```

Now that you know the different data types in Python and how to check the data type of a value you're working with, let's start automating simple arithmetic tasks.

USING LISTS TO STORE VALUES

So far we've used variables to hold a single value. A *list* is a type of variable that can hold multiple values, which is useful for automating repetitive tasks. To declare a list in Python, you simply create a name for the list, use the = command like you do with variables, and then enclose the items you want to place in the list in square brackets, [], separating each item using a comma, like this:

```
>>> a = [1,2,3]
>>> a
[1, 2, 3]
```

Often it's useful to create an empty list so you can add values, such as numbers, coordinates, and objects, to it later. To do this, just create the list as you would normally but without any values, as shown here:

```
>>> b = []
>>> b
[]
```

This creates an empty list called b, which you can fill with different values. Let's see how to add things to a list.

ADDING ITEMS TO A LIST

To add an item to a list, use the append() function, as shown here:

```
>>> b.append(4)
>>> b
[4]
```

First, type the name of the list (b) you want to add to, followed by a period, and then use append() to name the item you want to add inside parentheses. You can see the list now contains just the number 4.

You can also add items to lists that aren't empty, like this:

```
>>> b.append(5)
>>> b
[4, 5]
>>> b.append(True)
>>> b
[4, 5, True]
```

Items appended to an existing list appear at the end of the list. As you can see, your list doesn't have to be just numbers. Here, we append the Boolean value True to a list containing the numbers 4 and 5.

A single list can hold more than one data type, too. For example, you can add text as strings, as shown here:

```
>>> b.append("hello")
>>> b
[4, 5, True, 'hello']
```

To add a string, you need to include either double or single quotes around the text. Otherwise, Python looks for a variable named hello, which may or may not exist, thus causing an error or unexpected behavior. Now you have four items in list b: two numbers, a Boolean value, and a string.

OPERATING ON LISTS

Like on strings, you can use addition and multiplication operators on lists, but you can't simply add a number and a list. Instead, you have to append it using concatenation.

For example, you can add two lists together using the + operator, like this:

```
>>> c = [7,True]
>>> d = [8,'Python']
>>> c + d #adding two lists
[7, True, 8, 'Python']
```

We can also multiply a list by a number, like this:

```
>>> 2 * d #multiplying a list by a number
[8, 'Python', 8, 'Python']
```

As you can see, multiplying the number 2 by list d doubles the number of items in the original list.

But when we try to add a number and a list using the + operator, we get an error called a TypeError:

```
>>> d + 2 #you can't add a list and an integer
Traceback (most recent call last):
  File "<pyshell#22>", line 1, in <module>
    d + 2
TypeError: can only concatenate list (not "int") to list
```

This is because you can't add a number and a list using the addition symbol. Although you can add two lists together, append an item to a list, and even multiply a list by a number, you can concatenate a list only to another list.

REMOVING ITEMS FROM A LIST

Removing an item from a list is just as easy: you can use the remove() function with the item you want to remove as the argument, as shown next. Make sure to refer to the item you're removing exactly as it appears in the code; otherwise, Python won't understand what to delete.

```
>>> b = [4,5,True,'hello']
>>> b.remove(5)
>>> b
[4, True, 'hello']
```

In this example, b.remove(5) removes 5 from the list, but notice that the rest of the items stay in the same order. The fact that the order is maintained like this will become important later.

USING LISTS IN LOOPS

Often in math you need to apply the same action to multiple numbers. For example, an algebra book might define a function and ask you to plug a bunch of different numbers into the function. You can do this in Python by storing the numbers in a list and then using the for loop you learned about in Chapter 1 to perform the same action on each item in the list. Remember, when you perform an action repeatedly, it's known as *iterating*. The iterator is the variable i in for i in range(10), which we've used in previous programs, but it doesn't always have to be called i; it can be called anything you want, as in this example:

```
>>> a = [12,"apple",True,0.25]
>>> for thing in a:
        print(thing)

12
apple
True
0.25
```

Here, the iterator is called thing and it's applying the print() function to each item in the list a. Notice that the items are printed in order, with each item on a new line. To print everything on the same line, you need to add an end argument and an empty string to your print() function, like this:

```
>>> for thing in a:
        print(thing, end='')
12appleTrue0.25
```

This prints all the items on the same line, but all the values run together, making it hard to distinguish between them. The default value for the end argument is the line break, as you saw in the preceding example, but you can insert any character or punctuation you want by putting it in the quotes. Here I've added a comma instead:

```
>>> a = [12,"apple",True,0.25]
>>> for thing in a:
        print(thing, end=',')
12,apple,True,0.25,
```

Now each item is separated by a comma, which is much easier to read.

ACCESSING INDIVIDUAL ITEMS WITH LIST INDICES

You can refer to any element in a list by specifying the name of the list and then entering its index in square brackets. The *index* is an item's place or position number in the list. The first index of a list is 0. An index enables us to use a meaningful name to store a series of values and access them easily within our program. Try this code out in IDLE to see indices in action:

```
>>> name_list = ['Abe','Bob','Chloe','Daphne']
>>> score_list = [55,63,72,54]
>>> print(name_list[0], score_list[0])
Abe 55
```

The index can also be a variable or an iterator, as shown here:

```
>>> n = 2
>>> print(name_list[n], score_list[n+1])
Chloe 54
>>> for i in range(4):
        print(name_list[i], score_list[i])

Abe 55
Bob 63
Chloe 72
Daphne 54
```

ACCESSING INDEX AND VALUE WITH ENUMERATE()

To get both the index and the value of an item in a list, you can use a handy function called `enumerate()`. Here's how it works:

```
>>> name_list = ['Abe','Bob','Chloe','Daphne']
>>> for i, name in enumerate(name_list):
        print(name,"has index",i)

Abe has index 0
Bob has index 1
Chloe has index 2
Daphne has index 3
```

Here, `name` is the value of the item in the list and `i` is the index. The important thing to remember with `enumerate()` is that the index comes first, then the value. You'll see this later on when we put objects into a list and then access both an object and its exact place in the list.

INDICES START AT ZERO

In Chapter 1 you learned that the `range(n)` function generates a sequence of numbers starting with 0 and up to, but excluding, `n`. Similarly, list indices start at 0, not 1, so the index of the first element is `0`. Try the following to see how this works:

```
>>> b = [4,True,'hello']
>>> b[0]
4
>>> b[2]
'hello'
```

Here, we create a list called `b` and then ask Python to show us the item at index `0` in list `b`, which is the first position. We therefore get `4`. When we ask for the item in list `b` at position `2`, we get `'hello'`.

ACCESSING A RANGE OF LIST ITEMS

You can use the range (:) syntax inside the brackets to access a range of elements in a list. For example, to return everything from the second item of a list to the sixth, for example, use the following syntax:

```
>>> myList = [1,2,3,4,5,6,7]
>>> myList[1:6]
[2, 3, 4, 5, 6]
```

It's important to know that the `1:6` range syntax includes the *first* index in that range, `1`, but *excludes* the last index, `6`. That means the range `1:6` actually gives us the items with indexes `1` to `5`.

If you don't specify the ending index of the range, Python defaults to the length of the list. It returns all elements, from the first index to the end of the list, by default. For example, you can access everything from the second element of list `b` (index `1`) to the end of the list using the following syntax:

```
>>> b[1:]
[True, 'hello']
```

If you don't specify the beginning, Python defaults to the first item in the list, and it won't include the ending index, as shown here:

```
>>> b[:1]
[4]
```

In this example, `b[:1]` includes the first item (index `0`) but not the item with index `1`. One very useful thing to know is that you can access the last terms in a list even if you don't know how long it is by using negative numbers. To access the last item, you'd use `-1`, and to access the second-to-last item, you'd use `-2`, like this:

```
>>> b[-1]
'hello'
>>> b[-2]
True
```

This can be really useful when you are using lists made by other people or using really long lists where it's hard to keep track of all the index positions.

FINDING OUT THE INDEX OF AN ITEM

If you know that a certain value is in the list but don't know its index, you can find its location by giving the list name, followed by the `index` function, and placing the value you're searching for as its argument inside parentheses. In the shell, create list `c`, as shown here, and try the following:

```
>>> c = [1,2,3,'hello']
>>> c.index(1)
0
```

```
>>> c.index('hello')
3
>>> c.index(4)
Traceback (most recent call last):
  File "<pyshell#85>", line 1, in <module>
    b.index(4)
ValueError: 4 is not in list
```

You can see that asking for the value 1 returns the index 0, because it's the first item in the list. When you ask for the index of 'hello', you're told it's 3. That last attempt, however, results in an error message. As you can see from the last line in the error message, the cause of the error is that 4, the value we are looking for, is not in the list, so Python can't give us its index.

To check whether an item exists in a list, use the in keyword, like this:

```
>>> c = [1,2,3,'hello']
>>> 4 in c
False
>>> 3 in c
True
```

Here, Python returns True if an item is in the list and False if the item is not in the list.

STRINGS USE INDICES, TOO

Everything you've learned about list indices applies to strings, too. A string has a length, and all the characters in the string are indexed. Enter the following in the shell to see how this works:

```
>>> d = 'Python'
>>> len(d) #How many characters are in 'Python'?
6
>>> d[0]
'P'
>>> d[1]
'y'
>>> d[-1]
'n'
>>> d[2:]
'thon'
>>> d[:5]
'Pytho'
>>> d[1:4]
'yth'
```

Here, you can see that the string 'Python' is made of six characters. Each character has an index, which you can access using the same syntax you used for lists.

SUMMATION

When you're adding a bunch of numbers inside a loop, it's useful to keep track of the running total of those numbers. Keeping a running total like this is an important math concept called *summation*.

In math class you often see summation associated with a capital sigma, which is the Greek letter *S* (for sum). The notation looks like this:

$$\sum_{i=1}^{100} n$$

The summation notation means that you replace *n* with *i* starting at the minimum value (listed below the sigma) and going up to the maximum value (listed above the sigma). Unlike in Python's `range(n)`, the summation notation includes the maximum value.

CREATING THE RUNNING_SUM VARIABLE

To write a summation program in Python, we can create a variable called `running_sum` (`sum` is taken already as a built-in Python function). We set it to a value of zero to begin with and then increment the `running_sum` variable each time a value is added. For this we use the `+=` notation again. Enter the following in the shell to see an example:

```
>>> running_sum = 0
>>> running_sum += 3
>>> running_sum
3
>>> running_sum += 5
>>> running_sum
8
```

You learned how to use the `+=` command as a shortcut: using `running_sum += 3` is the same as `running_sum = running_sum + 3`. Let's increment the running sum by 3 a bunch of times to test it out. To do this, add the following code to the *arithmetic.py* program:

arithmetic.py

```
  running_sum = 0
❶ for i in range(10):
❷     running_sum += 3
  print(running_sum)
```

We first create a `running_sum` variable with the value `0` and then run the `for` loop 10 times using `range(10)` ❶. The indented content of the loop adds 3 to the value of `running_sum` on each run of the loop ❷. After the loop runs 10 times, Python jumps to the final line of code, which in this case is the `print` statement that displays the value of `running_sum` at the end of 10 loops.

From this, you might be able to figure out what the final sum is, and here's the output:

```
30
```

In other words, 10 multiplied by 3 is 30, so the output makes sense!

WRITING THE MYSUM() FUNCTION

Let's expand our running sum program into a function called `mySum()`, which takes an integer as a parameter and returns the sum of all the numbers from 1 up to the number specified, like this:

```
>>> mySum(10)
55
```

First, we declare the value of the running sum and then increment it in the loop:

arithmetic.py

```
def mySum(num):
    running_sum = 0
    for i in range(1,num+1):
        running_sum += i
    return running_sum
```

To define the `mySum()` function, we start the running sum off at 0. Then we set up a range of values for `i`, from 1 to `num`. Keep in mind that `range(1,num)` won't include `num` itself! Then we add `i` to the running sum after every loop. When the loop is finished, it should return the value of the running sum.

Run the function with a much larger number in the shell. It should be able to return the sum of all the numbers, from 1 to that number, in a flash:

```
>>> mySum(100)
5050
```

Pretty convenient! To solve for the sum of our more difficult sigma problem from earlier, simply change your loop to go from 0 to 20 (including 20) and add the square of `i` plus 1 every loop:

arithmetic.py

```
def mySum2(num):
    running_sum = 0
    for i in range(num+1):
        running_sum += i**2 + 1
    return running_sum
```

I changed the loop so it would start at 0, as the sigma notation indicates:

$$\sum_{i=0}^{20} n^2 + 1$$

When we run this, we get the following:

```
>>> mySum2(20)
2891
```

EXERCISE 2-1: FINDING THE SUM

Find the sum of all the numbers from 1 to 100. How about from 1 to 1,000? See a pattern?

FINDING THE AVERAGE OF A LIST OF NUMBERS

Now that you have a few new skills under your belt, let's improve our average function. We can write a function that uses lists to find the average of any list of numbers, without us having to specify how many there are.

In math class you learn that to find the average of a bunch of numbers, you divide the sum of those numbers by how many numbers there are. In Python you can use a function called `sum()` to add up all the numbers in a list, like this:

```
>>> sum([8,11,15])
34
```

Now we just have to find out the number of items in the list. In the `average()` function we wrote earlier in this chapter, we knew there were only two numbers. But what if there are more? Fortunately, we can use the `len()` function to count the number of items in a list. Here's an example:

```
>>> len([8,11,15])
3
```

As you can see, you simply enter the function and pass the list as the argument. This means that we can use both the `sum()` and `len()` functions to find the average of the items in the list by dividing the sum of the list by the length of the list. Using these built-in keywords, we can create a concise version of the average function, which would look something like this:

arithmetic.py

```
def average3(numList):
    return sum(numList)/len(numList)
```

When you call the function in the shell, you should get the following output:

```
>>> average3([8,11,15])
11.333333333333334
```

The good thing about this version of the average function is that it works for a short list of numbers as well as for a long one!

EXERCISE 2-2: FINDING THE AVERAGE

Find the average of the numbers in the list below:

```
d = [53, 28, 54, 84, 65, 60, 22, 93, 62, 27, 16, 25, 74, 42, 4, 42,
15, 96, 11, 70, 83, 97, 75]
```

SUMMARY

In this chapter you learned about data types like integers, floats, and Booleans. You learned to create a list, add and remove elements from a list, and find specific items in a list using indices. Then you learned how to use loops, lists, and variables to solve arithmetic problems, such as finding the average of a bunch of numbers and keeping a running sum.

In the next chapter you'll learn about conditionals, another important programming concept you'll need to learn to tackle the rest of this book.

3

GUESSING AND CHECKING WITH CONDITIONALS

"Put your dough into the oven when it is hot:
After making sure that it is in fact dough."
—Idries Shah, Learning How to Learn

In almost every program you write for this book, you're going to instruct the computer to make a decision. You can do this using an important programming tool called *conditionals.* In programming we can use conditional statements like "If this variable is more than 100, do this; otherwise, do that" to check whether certain conditions are met and then determine what to do based on the result. In fact, this is a very powerful method that we apply to big problems, and it's even at the heart of machine learning. At its most basic level, the program is guessing and then modifying its guesses based on feedback.

In this chapter you learn how to apply the guess-and-check method using Python to take user input and tell the program what to print depending on the input. You then use conditionals to compare different numerical values in different mathematical situations to make a turtle wander around the screen randomly. You also create a number-guessing game and use the same logic to find the square root of large numbers.

COMPARISON OPERATORS

As you learned in Chapter 2, True and False (which we capitalize in Python) are called Boolean values. Python returns Booleans when comparing two values, allowing you to use the result to decide what to do next. For example, we can use comparison operators like greater than (>) or less than (<) to compare two values, like this:

```
>>> 6 > 5
True
>>> 6 > 7
False
```

Here, we ask Python whether 6 is greater than 5, and Python returns True. Then we ask whether 6 is greater than 7, and Python returns False.

Recall that in Python we use one equal sign to assign a value to a variable. But checking for equality requires two equal signs (==), as shown here:

```
>>> 6 = 6
SyntaxError: can't assign to literal
>>> 6 == 6
True
```

As you can see, when we try to check using only one equal sign, we get a syntax error. We can also use comparison operators to compare variables:

```
>>> y = 3
>>> x = 4
>>> y > x
False
>>> y < 10
True
```

We set the variable y to contain 3, and then set the variable x to contain 4. Then we use those variables to ask whether y is greater than x, so Python returns False. Then we asked whether y is less than 10, which returns True. This is how Python makes comparisons.

MAKING DECISIONS WITH IF AND ELSE STATEMENTS

You can have your program make decisions about what code to run using if and else statements. For example, if the condition you set turns out to be True, the program runs one set of code. If the condition turns out to be False, you can write the program to do something else or even do nothing at all. Here's an example:

```
>>> y = 7
>>> if y > 5:
```

```
    print("yes!")
```

```
yes!
```

Here, we are saying, assign variable y the value 7. If the value of y is more than 5, print "yes!"; otherwise, do nothing.

You can also give your program alternative code to run using `else` and `elif`. Since we'll be writing some longer code, open a new Python file and save it as *conditionals.py*.

conditionals.py

```
y = 6
if y > 7:
    print("yes!")
else:
    print("no!")
```

In this example we're saying, if the value of y is more than 7, print "yes!"; otherwise, print "no!". Run this program, and it should print "no!" because 6 is not larger than 7.

You can add more alternatives using `elif`, which is short for "else if." You can have as many `elif` statements as you want. Here's a sample program with three `elif` statements:

conditionals.py

```
age = 50
if age < 10:
    print("What school do you go to?")
elif 11 < age < 20:
    print("You're cool!")
elif 20 <= age < 30:
    print("What job do you have?")
elif 30 <= age < 40:
    print("Are you married?")
else:
    print("Wow, you're old!")
```

This program runs different code depending on which of the specified ranges the value of `age` falls into. Notice you can use `<=` for "less than or equal to" and you can use compound inequalities like `if 11 < age < 20:` for "if age is between 11 and 20." For example, when `age = 50`, the output is the following string:

```
Wow, you're old!
```

Being able to have your programs make decisions quickly and automatically according to the conditions you define is an important aspect of programming!

USING CONDITIONALS TO FIND FACTORS

Now let's use what you've learned so far to factor a number! A *factor* is a number that divides evenly into another number; for example, 5 is a factor of 10 because we can divide 10 evenly by 5. In math class, we use factors to do everything from finding common denominators to determining whether a number is prime. But finding factors manually can be a tedious task involving a lot of trial and error, especially when you're working with bigger numbers. Let's see how to automate factoring using Python.

In Python you can use the modulo operator (`%`) to calculate the remainder when dividing two numbers. For example, if `a % b` equals zero, it means that `b` divides evenly into `a`. Here's an example of the modulo in action:

```
>>> 20 % 3
2
```

This shows that when you divide 20 by 3, you get a remainder of 2, which means that 3 is not a factor of 20. Let's try 5 instead:

```
>>> 20 % 5
0
```

Now we get a remainder of zero, so we know that 5 is a factor of 20.

WRITING THE FACTORS.PY PROGRAM

Let's use the modulo operator to write a function that takes a number and returns a list of that number's factors. Instead of just printing the factors, we'll put them in a list so we can use the factors list in another function later. Before we start writing this program, it's a good idea to lay out our plan. Here are the steps involved in the *factors.py* program:

1. Define the `factors` function, which takes a number as an argument.
2. Create an empty factors list to fill with factors.
3. Loop over all the numbers from 1 to the given number.
4. If any of these numbers divides evenly, add it to the factors list.
5. Return the list of factors at the end.

Listing 3-1 shows the `factors()` function. Enter this code into a new file in IDLE and save it as *factors.py*.

factors.py

```
def factors(num):
    '''returns a list of the factors of num'''
    factorList = []
    for i in range(1,num+1):
        if num % i == 0:
            factorList.append(i)
    return factorList
```

Listing 3-1: Writing the factors.py *program*

We first create an empty list called `factorList`, which we'll later fill with the factors as we find them. Then we start a loop, beginning with `1` (we can't divide by zero) and ending with `num + 1`, so that the loop will include num. Inside the loop we instruct the program to make a decision: if num is divisible by the current value of i (if the remainder is 0), then the program appends i to the factors list. Finally, we return the list of factors.

Now run *factors.py* by pressing the F5 key or by clicking **Run ▸ Run Module**, as shown in Figure 3-1.

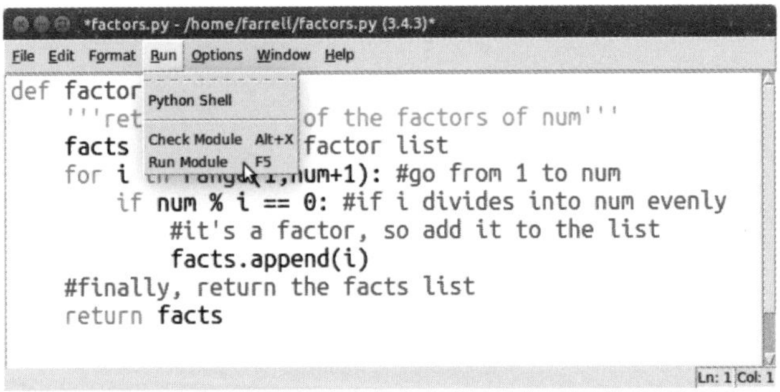

factors.py - /home/farrell/factors.py (3.4.3)

File Edit Format Run Options Window Help

Python Shell
Check Module Alt+X
Run Module F5

```
def factor
    '''ret            of the factors of num'''
    facts             factor list
    for i          ,num+1): #go from 1 to num
        if num % i == 0: #if i divides into num evenly
            #it's a factor, so add it to the list
            facts.append(i)
    #finally, return the facts list
    return facts
```

Ln: 1 Col: 1

Figure 3-1: Running the factors.py *module*

After running this module, you can use the factors function in the normal IDLE terminal by passing it a number you want to find the factors for, like this:

```
>>> factors(120)
[1, 2, 3, 4, 5, 6, 8, 10, 12, 15, 20, 24, 30, 40, 60, 120]
```

You found all the factors of 120 using the factors function! This is much easier and faster than using trial and error.

EXERCISE 3-1: FINDING THE FACTOR

The factors() function could come in handy for finding the greatest common factor (GCF) of two numbers. Write a function that will return the GCF of two numbers, like this:

```
>>> gcf(150,138)
6
```

THE WANDERING TURTLE

Now that you know how to instruct a program to make decisions automatically, let's explore how to let a program execute indefinitely! To start, we'll make a turtle walk around the screen and use conditionals to make it turn around if it goes beyond a certain point.

The turtle's window is a classic x-y grid whose x- and y-axes go from –300 to 300 by default. Let's limit the turtle's position to anywhere between –200 and 200 for *x* and *y*, as shown in Figure 3-2.

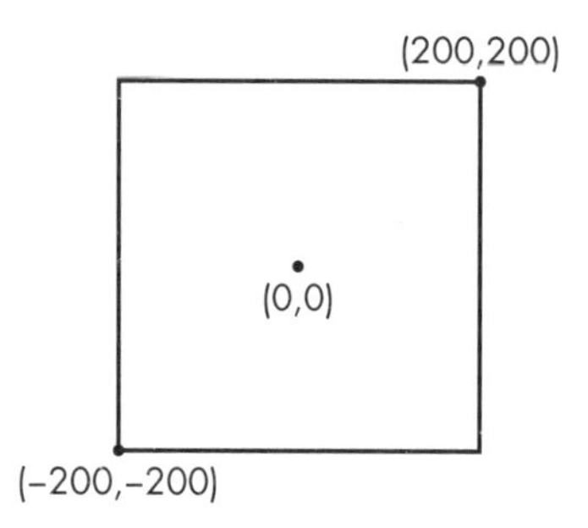

Figure 3-2: The rectangle of coordinates the turtle is limited to

Open a new Python file in IDLE and save it as *wander.py*. First, let's import the turtle module. To do so, add the following code:

```
from turtle import *
from random import randint
```

Note that we also need to import the `randint` function from the random module to generate random integers.

Writing the wander.py Program

Now let's create a function called `wander` to make the turtle wander around the screen, as shown in Listing 3-2. To do this, we use Python's infinite `while True` loop, which always evaluates to `True`. This will make the turtle wander around without stopping. To stop it, you can click the X on the turtle graphics window.

wander.py

```
speed(0)

def wander():
    while True:
        fd(3)
        if xcor() >= 200 or xcor() <= -200 or ycor()<= -200 or ycor() >= 200:
            lt(randint(90,180))

wander()
```

Listing 3-2: Writing the wander.py *program*

First, we set the turtle's speed to 0, which is the fastest, and then define the `wander()` function. Inside the function we use the infinite loop, so everything inside `while True` will execute forever. Then the turtle goes forward three steps (or 3 pixels) and evaluates its position using a conditional. The functions for the x-coordinate and y-coordinate of a turtle are `xcor()` and `ycor()`, respectively.

Using the `if` statement, we tell the program that if any one of the conditional statements is `True` (the turtle is outside the specified region), then make the turtle turn left a random number of degrees, between 90 and 180, to prevent it from straying. If the turtle is inside the rectangle, the conditional evaluates to False and no code is executed. Either way, the program returns to the top of the `while True` loop and does `fd(3)` again.

Running the wander.py Program

When you run the *wander.py* program, you should see something like Figure 3-3.

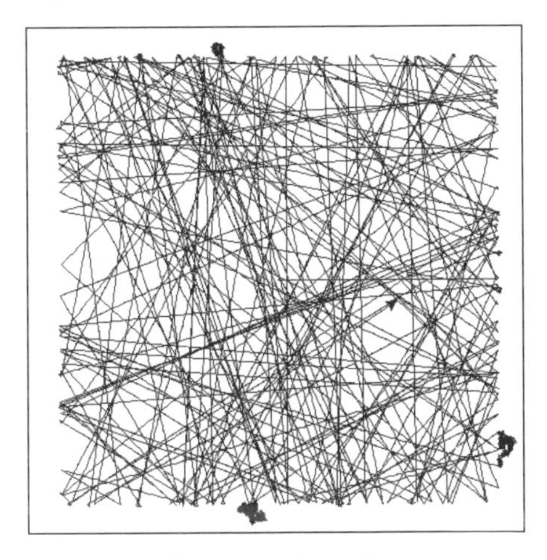

Figure 3-3: The output of wander.py

As you can see, the turtle walks in a straight line until its x-coordinate gets to 200. (The turtle always starts walking to the right in the positive x-direction.) Then it turns left a random number of degrees, between 90 and 180, and keeps walking straight again. Sometimes the turtle is able to walk outside of the boundary lines, because after the 90-degree turn it's still pointed off the screen and you'll see it turning around every loop, trying to get back into the rectangle. This causes the little blobs outside the rectangle you see in Figure 3-3.

CREATING A NUMBER-GUESSING GAME

You successfully used conditionals to create a turtle that seemed to make decisions on its own! Let's use conditionals to write an interactive number-guessing program that seems conscious. In this game I think of a number between 1 and 100, and you guess what the number is. How many guesses do you think you would need to guess my number correctly? To narrow down your options, after each incorrect guess, I tell you whether you should guess higher or lower. Fortunately, we can use the `average` function we wrote in Chapter 2 to make this task infinitely easier.

When you make an incorrect guess, your next guess should depend on whether your guess was too low or too high. For example, if your guess was too low, your next guess should be the middle number between your last guess and the maximum value the number can be. If your guess was too high, your next guess should be the middle number between your last guess and the minimum value the number can be.

This sounds like calculating the average of two numbers—good thing we have the average function! We'll use it to write the *numberGame.py* program, which makes smart guesses by narrowing down half the possible numbers every time. You'll be surprised how quickly you can hone in on the answer.

Let's take this one step at a time, starting with making a random number generator.

MAKING A RANDOM NUMBER GENERATOR

First, we need the computer to choose a number at random between 1 and 100. Create a new file in IDLE and save it as *numberGame.py*. Then enter the code in Listing 3-3.

numberGame.py

```
from random import randint

def numberGame():
    #choose a random number
    #between 1 and 100
    number = randint(1,100)
```

Listing 3-3: Writing the `numberGame()` *function*

Here, we import the `random` module and assign a random integer to a variable using the `randint()` function. Then we create a `number` variable that will hold a random number between 1 and 100, generated each time we call it.

TAKING USER INPUT

Now the program needs to ask the user for input so they can take a guess! Here's an example you can enter into the interactive shell to see how the `input()` function works:

```
>>> name = input("What's your name? ")
What's your name?
```

The program prints the text "What's your name?" in the shell, asking the user to input their name. The user types something, presses ENTER, and the program saves the input.

We can check whether Python saves the user input to the `name` variable, like so:

```
What's your name? Peter
>>> print(name)
Peter
```

When we ask the program to print `name`, it prints the user input that was saved in that variable (in this case, Peter).

We can create a function called greet() that we'll use later in our program:

```
def greet():
    name = input("What's your name? ")
    print("Hello, ",name)

greet()
```

The output will be the following:

```
>>>
What's your name? Al
Hello, Al
>>>
```

Try writing a short program that takes the user's name as input, and if they enter "Peter," it will print "That's my name, too!" If the name is not "Peter," it will just print "Hello" and the name.

CONVERTING USER INPUT TO INTEGERS

Now you know how to work with text that the user inputs, but we'll be taking in number inputs in our guessing game. In Chapter 2 you learned about basic data types, like integers and floats, that you can use to perform math operations. In Python, all input from users is always taken in as a *string*. This means that if we want numbers as inputs, we have to convert them to an integer data type so we can use them in operations.

To convert a string to an integer, we pass the input to int(), like this:

```
print("I'm thinking of a number between 1 and 100.")
guess = int(input("What's your guess? "))
```

Now whatever the user enters will be transformed into an integer that Python can operate on.

USING CONDITIONALS TO CHECK FOR A CORRECT GUESS

Now the *numberGame.py* program needs a way to check whether the number the user guessed is correct. If it is, we'll announce that the guess is right and the game is over. Otherwise, we tell the user whether they should guess higher or lower.

We use the if statement to compare the input to the content of number, and we use elif and else to decide what to do in each circumstance. Revise the existing code in *numberGame.py* to look like the code in Listing 3-4.

numberGame.py

```
from random import randint

def numberGame():
    #choose a random number
    #between 1 and 100
```

```
    number = randint(1,100)

    print("I'm thinking of a number between 1 and 100.")
    guess = int(input("What's your guess? "))

    if number == guess:
        print("That's correct! The number was", number)
    elif number > guess:
        print("Nope. Higher.")
    else:
        print("Nope. Lower.")

numberGame()
```

Listing 3-4: Checking for a correct guess

If the random number held in `number` is equal to the input stored in `guess`, we tell the user their guess was correct and print the random number. Otherwise, we tell the user whether they need to guess higher or lower. If the number they guessed is lower than the random number, we tell them to guess higher. If they guessed higher, we tell them to guess lower.

Here's an example of the output so far:

```
I'm thinking of a number between 1 and 100.
What's your guess? 50
Nope. Higher.
```

Pretty good, but currently our program ends here and doesn't let the user make any more guesses. We can use a loop to fix that.

USING A LOOP TO GUESS AGAIN!

To allow the user to guess again, we can make a loop so that the program keeps asking for more guesses until the user guesses correctly. We use the `while` loop to keep looping until `guess` is equal to `number`, and then the program will print a success message and break out of the loop. Replace the code in Listing 3-4 with the code in Listing 3-5.

numberGame.py

```
from random import randint

def numberGame():
    #choose a random number
    #between 1 and 100
    number = randint(1,100)

    print("I'm thinking of a number between 1 and 100.")
    guess = int(input("What's your guess? "))

    while guess:
        if number == guess:
            print("That's correct! The number was", number)
            break
        elif number > guess:
```

```
            print("Nope. Higher.")
        else:
            print("Nope. Lower.")
        guess = int(input("What's your guess? "))

numberGame()
```

Listing 3-5: Using a loop to allow the user to guess again

In this example, `while guess` means "while the variable `guess` contains a value." First, we check whether the random number it chose is equal to the guess. If it is, the program prints that the guess is correct and breaks out of the loop. If the number is greater than the guess, the program prompts the user to guess higher. Otherwise, it prints that the user needs to guess lower. Then it takes in the next guess and the loop starts over, allowing the user to guess as many times as needed to get the correct answer. Finally, after we're done defining the function, we write `numberGame()` to call the function to itself so the program can run it.

TIPS FOR GUESSING

Save the *numberGame.py* program and run it. Each time you make an incorrect guess, your next guess should be exactly halfway between your first guess and the closest end of the range. For example, if you start by guessing 50 and the program tells you to guess higher, your next guess would be halfway between 50 and 100 at the top of the range, so you'd guess 75.

This is the most efficient way to arrive at the correct number, because for each guess you're eliminating half the possible numbers, no matter whether the guess is too high or too low. Let's see how many guesses it takes to guess a number between 1 and 100. Figure 3-4 shows an example.

```
I'm thinking of a number between 1 and 100.
What's your guess? 50
Nope. Lower.
What's your guess? 25
Nope. Lower.
What's your guess? 12
Nope. Lower.
What's your guess? 6
Nope. Higher.
What's your guess? 9
Nope. Higher.
What's your guess? 10
That's correct! The number was 10
>>>
```

Figure 3-4: The output of the number-guessing game

This time it took six guesses.

Let's see how many times you can multiply 100 by a half before you get to a number below 1:

```
>>> 100*0.5
50.0
```

```
>>> 50*0.5
25.0
>>> 25*0.5
12.5
>>> 12.5*0.5
6.25
>>> 6.25*0.5
3.125
>>> 3.125*0.5
1.5625
>>> 1.5625*0.5
0.78125
```

It takes seven times to get to a number less than 1, so it makes sense that on average it takes around six or seven tries to guess a number between 1 and 100. This is the result of eliminating half the numbers in our range with every guess. This might not seem like a useful strategy for anything but number-guessing games, but we can use this exact idea to find a very accurate value for the square root of a number, which we'll do next.

FINDING SQUARE ROOTS

You can use the number-guessing game strategy to approximate square roots. As you know, some square roots can be whole numbers (the square root of 100 is 10, for example). But many more are *irrational numbers*, which are never-ending, never-repeating decimals. They come up a lot in coordinate geometry when you have to find the roots of polynomials.

So how could we possibly use the number-guessing game strategy to find an accurate value for a square root? You can simply use the averaging idea to calculate the square root, correct to eight or nine decimal places. In fact, your calculator or computer uses an iterative method like the number-guessing strategy to come up with square roots that are correct to 10 decimal places!

APPLYING THE NUMBER-GUESSING GAME LOGIC

For example, let's say you don't know the square root of 60. First, you narrow your options down to a range, like we did for the number-guessing game. You know that 7 squared is 49 and 8 squared is 64, so the square root of 60 must be between 7 and 8. Using the average() function, you can calculate the average of 7 and 8 to get 7.5, which is your first guess.

```
>>> average(7,8)
7.5
```

To check whether 7.5 is the correct guess, you can square 7.5 to see if it yields 60:

```
>>> 7.5**2
56.25
```

As you can see, 7.5 squared is 56.25. In our number-guessing game, we'd be told to guess higher since 56.25 is lower than 60.

Because we have to guess higher, we know the square root of 60 has to be between 7.5 and 8, so we average those and plug in the new guess, like so:

```
>>> average(7.5, 8)
7.75
```

Now we check the square of 7.75 to see if it's 60:

```
>>> 7.75**2
60.0625
```

Too high! So the square root must be between 7.5 and 7.75.

WRITING THE SQUAREROOT() FUNCTION

We can automate this process using the code in Listing 3-6. Open a new Python file and name it *squareRoot.py*.

squareRoot.py

```
def average(a,b):
    return (a + b)/2

def squareRoot(num,low,high):
    '''Finds the square root of num by
    playing the Number Guessing Game
    strategy by guessing over the
    range from "low" to "high"'''
    for i in range(20):
        guess = average(low,high)
        if guess**2 == num:
            print(guess)
        elif guess**2 > num: #"Guess lower."
            high = guess
        else: #"Guess higher."
            low = guess
    print(guess)

squareRoot(60,7,8)
```

Listing 3-6: Writing the `squareRoot()` function

Here, the `squareRoot()` function takes three parameters: `num` (the number we want the square root of), `low` (the lowest limit `num` can be), and `high` (the upper limit of `num`). If the number you guess squared is equal to `num`, we just print it and break out of the loop. This might happen for a whole number, but not for an irrational number. Remember, irrational numbers never end!

Next, the program checks whether the number you guess squared is greater than `num`, in which case you should guess lower. We shorten our range to go from `low` to the guess by replacing `high` with the guess. The only other possibility is if the guess is too low, in which case we shorten our range to go from the guess to `high` by replacing `low` with the guess.

The program keeps repeating that process as many times as we want (in this case, 20 times) and then prints the approximate square root. Keep in mind that any decimal, no matter how long, can only approximate an irrational number. But we can still get a very good approximation!

In the final line we call the `squareRoot()` function, giving it the number we want the square root of, and the low and high numbers in the range we know the square root has to be in. Our output should look like this:

```
7.745966911315918
```

We can find out how close our approximation is by squaring it:

```
>>> 7.745966911315918**2
60.00000339120106
```

That's pretty close to 60! Isn't it surprising that we can calculate an irrational number so accurately just by guessing and averaging?

EXERCISE 3-2: FINDING THE SQUARE ROOT

Find the square root of these numbers:

- 200
- 1000
- 50000 (Hint: you know the square root has to be somewhere between 1 and 500, right?)

SUMMARY

In this chapter, you learned about some handy tools like arithmetic operators, lists, inputs, and Booleans, as well as a crucial programming concept called conditionals. The idea that we can get the computer to compare values and make choices for us automatically, instantly, and repeatedly is extremely powerful. Every programming language has a way to do this, and in Python we use `if`, `elif`, and `else` statements. As you'll see throughout this book, you'll build on these tools to tackle meatier tasks to explore math.

In the next chapter, you'll practice the tools you learned so far to solve algebra problems quickly and efficiently. You'll use the number-guessing strategy to solve complicated algebraic equations that have more than one solution! And you'll write a graphing program so you can better estimate the solutions to equations and make your math explorations more visual!

PART 2

RIDING INTO MATH TERRITORY

TRANSFORMING AND STORING NUMBERS WITH ALGEBRA

"Mathematics may be defined as the subject in which we never know what we are talking about, nor whether what we are saying is true."
—Bertrand Russell

If you learned algebra in school, you're probably familiar with the idea of replacing numbers with letters. For example, you can write $2x$ where x is a placeholder that can represent any number. So $2x$ represents the idea of multiplying two by some unknown number. In math class, variables become "mystery numbers" and you're required to find what numbers the letters represent. Figure 4-1 shows a student's cheeky response to the problem "Find x."

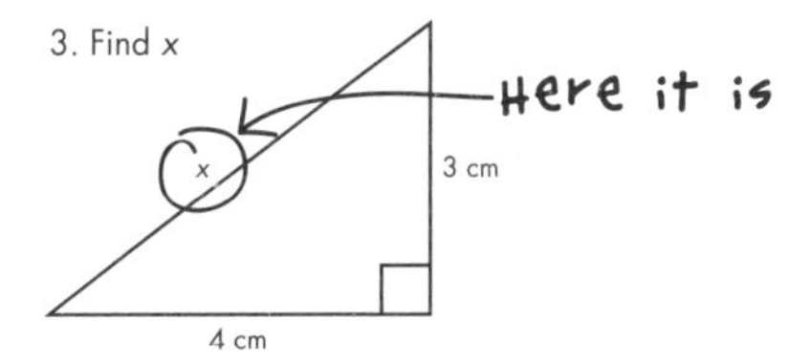

Figure 4-1: Locating the x variable instead of solving for its value

As you can see, this student has located the variable *x* in the diagram instead of *solving* for its value. Algebra class is all about solving equations like this: solve $2x + 5 = 13$. In this context, "to solve" means to figure out which number, when you replace *x* with that number, makes the equation true. You can solve algebra problems by balancing equations, which requires a lot of rules you have to memorize and follow.

Using letters as placeholders in this way is just like using variables in Python. In fact, you already learned how to use variables to store and calculate numerical values in previous chapters. The important skill math students should learn is not solving for variables but rather *using* variables. In fact, solving equations by hand is only of limited value. In this chapter you use variables to write programs that find unknown values quickly and automatically without having to balance equations! You also learn to use a programming environment called Processing to graph functions to help you explore algebra visually.

SOLVING FIRST-DEGREE EQUATIONS

One way to solve a simple equation like $2x + 5 = 13$ with programming is by using *brute force* (that is, plugging in random numbers until we find the right one). For this particular equation we need to find a number, *x*, that when we multiply it by 2 and then add 5, returns 13. I'll make an educated guess that *x* is a value between –100 and 100, since we're working with mostly double-digit numbers or lower.

This means that we can write a program that plugs all the integers between –100 and 100 into the equation, checks the output, and prints the number that makes the equation true. Open a new file in IDLE, save it as *plug.py*, and enter the code in Listing 4-1 to see such a program in action.

```
def plug():
❶ x = -100 #start at -100
   while x < 100: #go up to 100
    ❷ if 2*x + 5 == 13: #if it makes the equation true
           print("x =",x) #print it out
    ❸ x += 1 #make x go up by 1 to test the next number

plug() #run the plug function
```

Listing 4-1: Brute-force program that plugs in numbers to see which one satisfies the equation

Here, we define the `plug()` function and initialize the `x` variable at `-100` ❶. On the next line we start a `while` loop that repeats until `x` equals 100, which is the upper limit of the range we set. We then multiply `x` by 2 and add 5 ❷. If the output equals 13, we tell the program to print the number, because that's the solution. If the output does not equal 13, we tell the program to keep going through the code.

The loop then starts over, and the program tests the next number, which we get by incrementing `x` by 1 ❸. We continue the loop until we hit a match. Be sure to include the last line, which makes the program run the `plug()` function we just defined; if you don't, your program won't do anything! The output should be this:

```
x = 4
```

Using the guess-and-check method is a perfectly valid way to solve this problem. Plugging in all the digits by hand can be laborious, but using Python makes it a cinch! If you suspect the solution isn't an integer, you might have to increment by smaller numbers by changing the line at ❸ to `x += .25` or some other decimal value.

FINDING THE FORMULA FOR FIRST-DEGREE EQUATIONS

Another way to solve an equation like $2x + 5 = 13$ is to find a general formula for this type of equation. We can then use this formula to write a program in Python. You might recall from math class that the equation $2x + 5 = 13$ is an example of a *first-degree equation*, because the highest exponent a variable has in this equation is 1. And you probably know that a number raised to the first power equals the number itself.

In fact, all first-degree equations fit into this general formula: $ax + b = cx + d$, where a, b, c, and d represent different numbers. Here are some examples of other first-degree equations:

$$3x - 5 = 22$$

$$4x - 12 = 2x - 9$$

$$\frac{1}{2}x + \frac{2}{3} = \frac{1}{5}x + \frac{7}{8}$$

On each side of the equal sign, you can see an x term and a *constant*, which is a number with no x attached to it. The number that precedes the x variable is called a *coefficient*. For example, the coefficient of $3x$ is 3.

But sometimes there's no x term at all on one side of the equation, which means that the coefficient of that x is zero. You can see this in the first example, $3x - 5 = 22$, where 22 is the only term on the right side of the equal sign:

$$ax + b = cx + d$$

$$3x - 5 = 0 + 22$$

Using the general formula, you can see that a = 3, b = –5, and d = 22. The only thing that seems to be missing is the value of c. But it's not actually missing. In fact, the fact that there's nothing there means $cx = 0$, which means that c must equal zero.

Now let's use a little algebra to solve the equation $ax + b = cx + d$ for x. If we can find what x is in the formula, we can use it to solve virtually all equations of this form.

To solve this equation, we first get all the x's on one side of the equal sign by subtracting cx and b from both sides of the equation, like this:

$$ax - cx = d - b$$

Then we can factor out the x from ax and cx:

$$x(a - c) = d - b$$

Finally, divide both sides by a – c to isolate x, which gives us the value of x in terms of a, b, c, and d:

$$x = \frac{d - b}{a - c}$$

Now you can use this general equation to solve for any variable x when the equation is a first-degree equation and all coefficients (a, b, c, and d) are known. Let's use this to write a Python program that can solve first-degree algebraic equations for us.

WRITING THE EQUATION() FUNCTION

To write a program that will take the four coefficients of the general equation and print out the solution for x, open a new Python file in IDLE. Save it as *algebra.py*. We'll write a function that takes the four numbers a, b, c, and d as parameters and plug them into the formula (see Listing 4-2).

```
def equation(a,b,c,d):
    ''''solves equations of the
    form ax + b = cx + d''''
    return (d - b)/(a - c)
```

Listing 4-2: Using programming to solve for x

Recall that the general formula of a first-degree equation is this:

$$x = \frac{d - b}{a - c}$$

This means that for any equation with the form ax + b = cx + d, if we take the coefficients and plug them into this formula, we can calculate the x value. First, we set the `equation()` function to take the four coefficients as its parameters. Then we use the expression `(d - b)/(a - c)` to represent the general equation.

Now let's test our program with an equation you've already seen: $2x + 5 = 13$. Open the Python shell, type the following code at the >>> prompt, and press ENTER:

```
>>> equation(2,5,0,13)
4.0
```

If you input the coefficients of this equation into the function, you get 4 as the solution. You can confirm that it's correct by plugging in 4 in place of x. It works!

EXERCISE 4-1: SOLVING MORE EQUATIONS FOR X

Solve $12x + 18 = -34x + 67$ using the program you wrote in Listing 4-2.

USING PRINT() INSTEAD OF RETURN

In Listing 4-2, we used `return` instead of `print()` to display our results. This is because `return` gives us our result as a number that we can assign to a variable and then use again. Listing 4-3 shows what would happen if we used `print()` instead of `return` to find *x*:

```
def equation(a,b,c,d):
    ''''solves equations of the
    form ax + b = cx + d''''
    print((d - b)/(a - c))
```

Listing 4-3: Using `print()` doesn't let us save the output

When you run this, you get the same output:

```
>>> x = equation(2,5,0,13)
4.0
>>> print(x)
None
```

But when you try to call the `x` value using `print()`, the program doesn't recognize your command because it hasn't saved the result. As you can see, `return` can be more useful in programming because it lets you save the output of a function so you can apply it elsewhere. This is why we used `return` in Listing 4-2.

To see how you can work with the returned output, use the equation $12x + 18 = -34x + 67$ from Exercise 4-1 and assign the result to the `x` variable, as shown here:

```
>>> x = equation(12,18,-34,67)
>>> x
1.065217391304348
```

First, we pass the coefficients and constants of our equation to the `equation()` function so that it solves the equation for us and assigns the solution to the variable `x`. Then we can simply enter **`x`** to see its value. Now that the variable `x` stores the solution, we can plug it back into the equation to check that it's the correct answer.

Enter the following to find out what $12x + 18$, the left side of the equation, evaluates to:

```
>>> 12*x + 18
30.782608695652176
```

We get `30.782608695652176`. Now enter the following to do the same for $-34x + 67$, the right side of the equation:

```
>>> -34*x + 67
30.782608695652172
```

Except for a slight rounding discrepancy at the 15th decimal place, you can see that both sides of the equation evaluate to around 30.782608. So we can be confident that 1.065217391304348 is indeed the correct solution for *x*! Good thing we returned the solution and saved the value instead of just printing it out once. After all, who wants to type in a number like 1.065217391304348 again and again?

EXERCISE 4-2: FRACTIONS AS COEFFICIENTS

Use the `equation()` function to solve the last, most sinister-looking equation you saw on page 55:

$$\frac{1}{2}x + \frac{2}{3} = \frac{1}{5}x + \frac{7}{8}$$

SOLVING HIGHER-DEGREE EQUATIONS

Now that you know how to write programs that solve for unknown values in first-degree equations, let's try something harder. For example, things get a little more complicated when an equation has a term raised to the second degree, like $x^2 + 3x - 10 = 0$. These are called *quadratic equations*, and their general form looks like $ax^2 + bx + c = 0$, where a, b, and c can be any number: positive or negative, whole numbers, fractions, or decimals. The only exception is that a can't be 0 because that would make this a first-degree equation. Unlike first-degree equations, which have one solution, quadratic equations have two possible solutions.

To solve an equation with a squared term, you can use the *quadratic formula*, which is what you get when you isolate *x* by balancing the equation $ax^2 + bx + c = 0$:

$$x = \frac{-b \pm \sqrt{b^2 - 4ac}}{2a}$$

The quadratic formula is a very powerful tool for solving equations, because no matter what a, b, and c are in $ax^2 + bx + c = 0$, you can just plug them in to the formula and use basic arithmetic to find your solutions.

We know that the coefficients of $x^2 + 3x - 10 = 0$ are 1, 3, and –10. When we plug those in to the formula, we get

$$x = \frac{-3 \pm \sqrt{3^2 - 4(1)(-10)}}{2(1)}$$

Isolate x and this simplifies to

$$x = \frac{-3 \pm \sqrt{49}}{2} = \frac{-3 \pm 7}{2}$$

There are two solutions:

$$x = \frac{-3 + 7}{2}$$

which is equal to 2, and

$$x = \frac{-3 - 7}{2}$$

which is equal to –5.

We can see that replacing x in the quadratic formula with either of these solutions makes the equation true:

$$(2)^2 + 3(2) - 10 = 4 + 6 - 10 = 0$$
$$(-5)^2 + 3(-5) - 10 = 25 - 15 - 10 = 0$$

Next, we'll write a function that uses this formula to return two solutions for any quadratic equation.

USING QUAD() TO SOLVE QUADRATIC EQUATIONS

Let's say we want to use Python to solve the following quadratic equation:

$$2x^2 + 7x - 15 = 0$$

To do this, we'll write a function called `quad()` that takes the three coefficients (a, b, and c) and returns two solutions. But before we do anything, we need to import the `sqrt` method from the math module. The `sqrt` method allows us to find the square root of a number in Python, just like a square root button on a calculator. It works great for positive numbers, but if you try finding the square root of a negative number, you'll see an error like this:

```
>>> from math import sqrt
>>> sqrt(-4)
Traceback (most recent call last):
  File "<pyshell#11>", line 1, in <module>
    sqrt(-4)
ValueError: math domain error
```

Open a new Python file in IDLE and name it *polynomials.py*. Add the following line to the top of your file to import the sqrt function from the math module:

```
from math import sqrt
```

Then enter the code in Listing 4-4 to create the quad() function.

```
def quad(a,b,c):
    ''''Returns the solutions of an equation
    of the form a*x**2 + b*x + c = 0''''
    x1 = (-b + sqrt(b**2 - 4*a*c))/(2*a)
    x2 = (-b - sqrt(b**2 - 4*a*c))/(2*a)
    return x1,x2
```

Listing 4-4: Using the quadratic formula to solve an equation

The quad() function takes the numbers a, b, and c as parameters and plugs them in to the quadratic formula. We use x1 to assign the result of (the first solution), and x2 will store the value of (the second solution).

Now, let's test this program to solve for x in $2x^2 + 7x - 15 = 0$. Plugging in the numbers 2, 7, and –15 for a, b, and c should return the following output:

```
>>> quad(2,7,-15)
(1.5, -5.0)
```

As you can see, the two solutions for x are 1.5 and –5, which means both values should satisfy the equation $2x^2 + 7x - 15 = 0$. To test this, replace all the x variables in the original equation $2x^2 + 7x - 15 = 0$ with 1.5, the first solution, and then with –5, the second solution, as shown here:

```
>>> 2*1.5**2 + 7*1.5 - 15
0.0
>>> 2*(-5)**2 + 7*(-5) - 15
0
```

Success! This confirms that both values work in the original equation. You can use the equation() and quad() functions any time in the future. Now that you've learned to write functions to solve first-degree and second-degree equations, let's discuss how to solve even higher-degree equations!

USING PLUG() TO SOLVE A CUBIC EQUATION

In algebra class, students are often asked to solve a *cubic equation* like $6x^3 + 31x^2 + 3x - 10 = 0$, which has a term raised to the third degree. We can tweak the plug() function we wrote in Listing 4-1 to solve this cubic equation using the brute-force method. Enter the code shown in Listing 4-5 into IDLE to see this in action.

plug.py

```
def g(x):
    return 6*x**3 + 31*x**2 + 3*x - 10

def plug():
    x = -100
    while x < 100:
        if g(x) == 0:
            print("x =",x)
        x += 1
    print("done.")
```

Listing 4-5: Using plug() to solve a cubic equation

First, we define `g(x)` to be a function that evaluates the expression `6*x**3 + 31*x**2 + 3*x - 10`, the left side of our cubic equation. Then we tell the program to plug all numbers between –100 and 100 into the `g(x)` function we just defined. If the program finds a number that makes `g(x)` equal zero, then it has found the solution and prints it for the user.

When you call `plug()`, you should see the following output:

```
>>> plug()
x = -5
done.
```

This gives you –5 as the solution, but as you might suspect from working with quadratic equations previously, the x^3 term means there could be as many as three solutions to this equation. As you can see, you can brute-force your way to a solution like this, but you won't be able to determine whether other solutions exist or what they are. Fortunately, there's a way to see all the possible inputs and corresponding outputs of a function; it's called *graphing*.

SOLVING EQUATIONS GRAPHICALLY

In this section, we'll use a nifty tool called Processing to graph higher-degree equations. This tool will help us find solutions to higher-degree equations in a fun and visual way! If you haven't already installed Processing, follow the instructions in "Installing Processing" on page xxiii.

GETTING STARTED WITH PROCESSING

Processing is a programming environment and a graphics library that makes it easy to visualize your code. You can see the cool, dynamic, interactive art you can make with Processing in the examples page at *https://processing.org/examples/*. You can think of Processing as a sketchbook for your programming ideas. In fact, each Processing program you create is called a *sketch*. Figure 4-2 shows what a short Processing sketch in Python mode looks like.

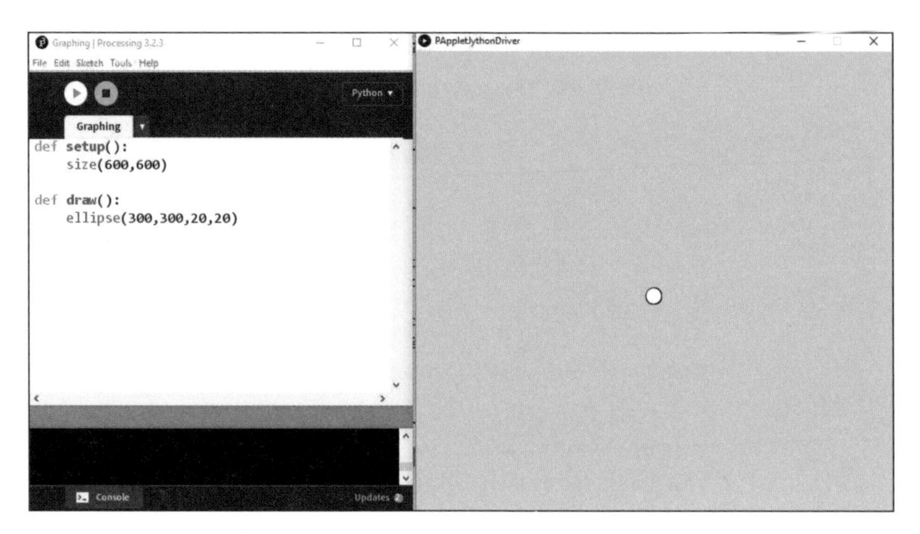

Figure 4-2: Example of a Processing sketch

As you can see, there's a programming environment where you enter code and a separate *display window* that shows the visualization of the code. This is a sketch of a simple program that creates a small circle. Every one of the Processing sketches we'll create will contain two of Processing's built-in functions: `setup()` and `draw()`. The code we put in the `setup()` function will be run once, when you click the play button, the arrow at the top left of the interface. Whatever we put in `draw()` will be repeated as an infinite loop until you click the stop button next to the play button.

In Figure 4-2 you can see in the `setup()` function we declared the size of the display screen to be 600 pixels by 600 pixels using the `size()` function. In the `draw()` function we told the program to draw a circle using the `ellipse()` function. Where? How big? We have to tell the `ellipse()` function four numbers: the x-coordinate of the ellipse, its y-coordinate, its width, and its height.

Notice the circle shows up in the middle of the screen, which in math class is the *origin* (0,0). But in Processing and in many other graphics libraries, (0,0) is in the top left corner of the screen. So to put a circle in the middle, I had to divide the length of the window (600) and the width of the window (600) in half. So its location is (300,300), and not (0,0).

Processing has a number of functions, like `ellipse()`, that make it easy to draw shapes. To see the full list, look at the reference pages at *https://processing.org/reference/* to find functions for drawing ellipses, triangles, rectangles, arcs, and much more. We'll explore drawing shapes with Processing in more detail in the next chapter.

NOTE *The code colors in Processing appear different from those used in IDLE. For example, you can see that* `def` *appears green in Processing in Figure 4-2, whereas it is orange in IDLE.*

CREATING YOUR OWN GRAPHING TOOL

Now that you've downloaded Processing, let's use it to create a graphing tool that allows us to see how many solutions an equation has. First, we create a grid of blue lines that looks like graphing paper. Then, we create the x- and y-axes using black lines.

Setting Graph Dimensions

In order to make a grid for our graphing tool, we first need to set the dimensions of the display window. In Processing, you can use the `size()` function to indicate the width and height of the screen in pixels. The default screen size is 600 pixels by 600 pixels, but for our graphing tool we'll create a graph that includes x- and y-values ranging from –10 to 10.

Open a new file in Processing and save it as *grid.pyde*. Make sure you're in Python mode. Enter the code in Listing 4-6 to declare the range of x- and y-values we're interested in displaying for our graph.

grid.pyde

```
#set the range of x-values
xmin = -10
xmax = 10

#range of y-values
ymin = -10
ymax = 10

#calculate the range
rangex = xmax - xmin
rangey = ymax - ymin

def setup():
    size(600,600)
```

Listing 4-6: Setting the range of x- and y-values for the graph

In Listing 4-6 we create two variables, `xmin` and `xmax`, for the minimum and maximum x-values in our grid, then we repeat the process for the y-values. Next we declare `rangex` for the x-range and `rangey` variable for the y-range. We calculate the value of `rangex` by subtracting `xmin` from `xmax` and do the same for the y-values.

Because we don't need a graph that's 600 units by 600 units, we need to scale the coordinates down by multiplying the x- and y-coordinates by scale factors. When graphing we have to remember to multiply all our x-coordinates and y-coordinates by these scale factors; otherwise, they won't show up correctly on the screen. To do this, update the existing code in the `setup()` function with the lines of code in Listing 4-7.

grid.pyde

```
def setup()
    global xscl, yscl
    size(600,600)
```

```
    xscl = width / rangex
    yscl = -height / rangey
```

Listing 4-7: Scaling coordinates using scale factors

First, we declare the global variables `xscl` and `yscl`, which we'll use to scale our screen. `xscl` and `yscl` stand for the x-scale factor and y-scale factor, respectively. For example, the x-scale factor would be 1 if we want our x-range to be 600 pixels, or the full width of the screen. But if we want our screen to be between –300 and 300, the x-scale factor would be 2, which we get by dividing the `width` (600) by the `rangex` (300).

In our case, we can calculate the scale factor by dividing 600 by the x-range, which is 20 (–10 to 10). So the scale factor has to be 30. From now on, we need to scale up all of our x- and y-coordinates by a factor of 30 so that they show on the screen. The good news is that the computer will do all the dividing and scaling for us. We just have to remember to use `xscl` and `yscl` when graphing!

Drawing a Grid

Now that we've set the proper dimensions for our graph, we can draw grid lines like the ones you see on graphing paper. Everything in the `setup()` function will be run once. Then we create an infinite loop with a function called `draw()`. `Setup()` and `draw()` are built-in Processing functions, and you can't change their names if you want the sketch to run. Add the code in Listing 4-8 to create the `draw()` function.

grid.pyde

```
#set the range of x-values
xmin = -10
xmax = 10

#range of y-values
ymin = -10
ymax = 10

#calculate the range
rangex = xmax - xmin
rangey = ymax - ymin

def setup():
    global xscl, yscl
    size(600,600)
    xscl = width / rangex
    yscl = height / rangey

def draw():
    global xscl, yscl
    background(255) #white
    translate(width/2,height/2)
    #cyan lines
    strokeWeight(1)
```

```
stroke(0,255,255)
for i in range(xmin,xmax + 1):
    line(i*xscl,ymin*yscl,i*xscl,ymax*yscl)
    line(xmin*xscl,i*yscl,xmax*xscl,i*yscl)
```

Listing 4-8: Creating blue grid lines for the graph

First, we use `global xscl, yscl` to tell Python we're not creating new variables but just using the global ones we already created. Then we set the background color to white using the value 255. We use Processing's `translate()` function to move shapes up and down, or left and right. The code `translate(width/2,height/2)` will move the origin (where x and y are both 0) from the top left to the center of the screen. Then we set the thickness of the lines with `strokeWeight`, where `1` is the thinnest. You can make them thicker if you want by using higher numbers. You can also change the color of the lines using `stroke`. Here, we're using cyan ("sky blue"), whose RGB value is (0,255,255), which means no red values, maximum green, and maximum blue.

After that, we use a `for` loop to avoid having to type 40 lines of code to draw 40 blue lines. We want the blue lines to go from `xmin` to `xmax`, including `xmax`, because that's how wide our graph should be.

RGB VALUES

An RGB value is a mixture of red, green, and blue, in that order. The values range from 0 to 255. For example, (255,0,0) means "maximum red, no green, no blue." Yellow is a mixture of red and green only, and cyan ("sky blue") is a mixture of green and blue only.

(255,0,0) (255,255,0) (0,255,0) (0,255,255) (0,0,255)

Other colors are a mixture of different levels of red, green, and blue:

(255,0,255) (128,0,128) (255,140,0) (102,51,0) (250,128,114)

You can do a web search for "RGB Tables" to get RGB values for many more colors!

In Processing, you can draw a line by declaring four numbers: the x- and y-coordinates of the beginning and endpoints of the line. The vertical lines would look something like this:

```
line(-10,-10, -10,10)
line(-9,-10, -9,10)
line(-8,-10, -8,10)
```

But because `range(x)` doesn't include `x` (as you learned previously), our `for` loop needs to go from `xmin` to `xmax + 1` to include `xmax`.

Similarly, the horizontal lines would go like this:

```
line(-10,-10, 10,-10)
line(-10,-9, 10,-9)
line(-10,-8, 10,-8)
```

This time, you can see that the y-values are –10, –9, –8 and so on, whereas the x-values stay constant at –10 and 10, which are `xmin` and `xmax`. Let's add another loop to go from `ymin` to `ymax`:

```
for i in range(xmin,xmax+1):
    line(i,ymin,i,ymax)
for i in range(ymin,ymax+1):
    line(xmin,i,xmax,i)
```

If you graphed this right, you would now see a tiny splotch in the middle of the screen because the x- and y-coordinates go from –10 to 10, but the screen goes from 0 to 600 by default. This is because we haven't multiplied all our x- and y-coordinates by their scale factor yet! To display the grid properly, update your code as follows:

```
for i in range(xmin,xmax+1):
    line(i*xscl,ymin*yscl,i*xscl,ymax*yscl)
for i in range(ymin,ymax+1):
    line(xmin*xscl,i*yscl,xmax*xscl,i*yscl)
```

Now you're ready to create the x- and y-axes.

Creating the X- and Y-Axes

To add the two black lines for the x- and y-axes, we first set the stroke color to black by calling the `stroke()` function (with 0 being black and 255 being white). Then we draw a vertical line from (0,–10) to (0,10) and a horizontal line from (–10,0) to (10,0). Don't forget to multiply the values by their respective scale factors, unless they're 0, in which case multiplying them wouldn't change them anyway.

Listing 4-9 shows the complete code for creating the grid.

grid.pyde

```
#cyan lines
strokeWeight(1)
stroke(0,255,255)
```

```
    for i in range(xmin,xmax+1):
        line(i*xscl,ymin*yscl,i*xscl,ymax*yscl)
    for i in range(ymin,ymax+1):
        line(xmin*xscl,i*yscl,xmax*xscl,i*yscl)
    stroke(0) #black axes
    line(0,ymin*yscl,0,ymax*yscl)
    line(xmin*xscl,0,xmax*xscl,0)
```

Listing 4-9: Creating the grid lines

When you click **Run**, you should get a nice grid, like in Figure 4-3.

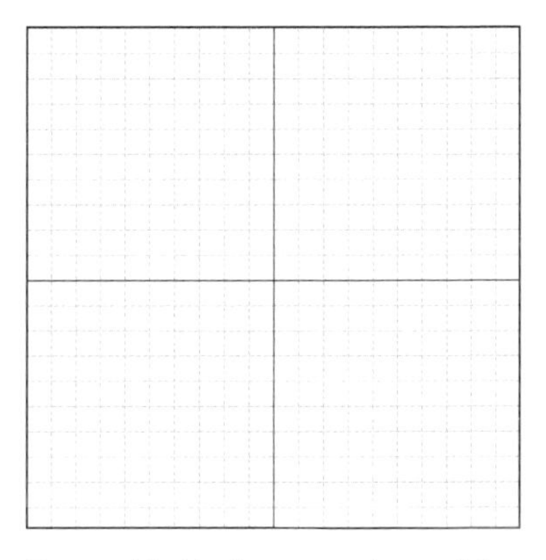

Figure 4-3: You've created a grid for graphing—and you only have to do it once!

This looks done, but if we try to put a point (a tiny ellipse, actually) at (3,6), we see a problem. Add the following code to the end of the `draw()` function:

grid.pyde

```
#test with a circle
fill(0)
ellipse(3*xscl,6*yscl,10,10)
```

When you run this, you'll see the output in Figure 4-4.

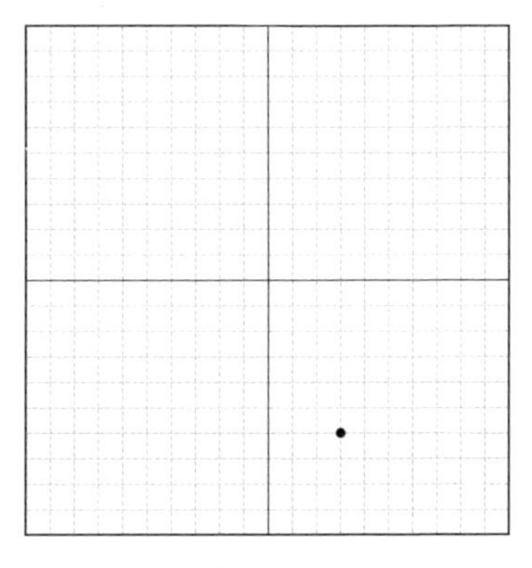

Figure 4-4: Checking our graphing program. Almost there!

As you can see, the point ends up on (3,–6) instead of at (3,6). Our graph is upside-down! To fix this, we can add a negative sign to the y-scale factor in the `setup()` function to flip it over:

```
yscl = -height/rangey
```

Now, you should see the point at the correct location, like in Figure 4-5.

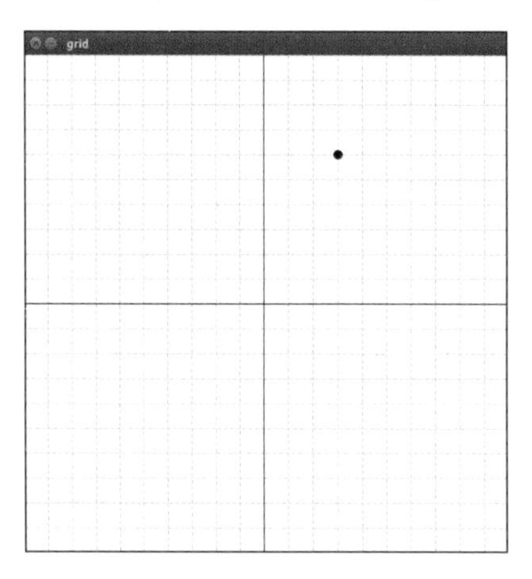

Figure 4-5: The grapher is working properly!

Now that we've written the graphing tool, let's put it into a function so we can reuse it whenever we need to graph an equation.

Writing the grid() Function

To keep our code organized, we'll separate all the code that makes the grid and place it in its own function, which we'll call `grid()`. Then we'll call the `grid()` function in the `draw()` function like in Listing 4-10.

grid.pyde

```
def draw():
    global xscl, yscl
    background(255)
    translate(width/2,height/2)
    grid(xscl,yscl) #draw the grid

def grid(xscl,yscl):
    #Draws a grid for graphing
    #cyan lines
    strokeWeight(1)
    stroke(0,255,255)
    for i in range(xmin,xmax+1):
        line(i*xscl,ymin*yscl,i*xscl,ymax*yscl)
    for i in range(ymin,ymax+1):
        line(xmin*xscl,i*yscl,xmax*xscl,i*yscl)
    stroke(0) #black axes
```

```
    line(0,ymin*yscl,0,ymax*yscl)
    line(xmin*xscl,0,xmax*xscl,0)
```

Listing 4-10: Moving all the grid code into a separate function

In programming we often organize our code into functions. Notice in Listing 4-10 we can easily see what we're executing in our `draw()` function. Now we're ready to solve our cubic equation, $6x^3 + 31x^2 + 3x - 10 = 0$.

GRAPHING AN EQUATION

Plotting graphs is a fun and visual way to find solutions of polynomials that have more than one potential solution for x. But before we try to graph a complicated equation like $6x^3 + 31x^2 + 3x - 10 = 0$, let's plot a simple parabola.

Plotting Points

Add this function after the `draw()` function from Listing 4-10:

grid.pyde

```
def f(x):
    return x**2
```

This defines the function we're calling `f(x)`. We're telling Python what to do with the number x to produce the output of the function. In this case, we're telling it to square the number x and return the output. Math classes have traditionally called functions `f(x)`, `g(x)`, `h(x)` and so on. Using a programming language, you can call functions whatever you like! We could have given this function a descriptive name like `parabola(x)`, but since `f(x)` is commonly used, we'll stick to that for now.

This is a simple parabola that we'll graph before getting into more complicated functions. All the points on this curve are simply the values for `x` and its corresponding y-value. We could use a loop and draw small ellipses for points at all the whole-number values for `x`, but that would look like an unconnected group of points, as in Figure 4-6.

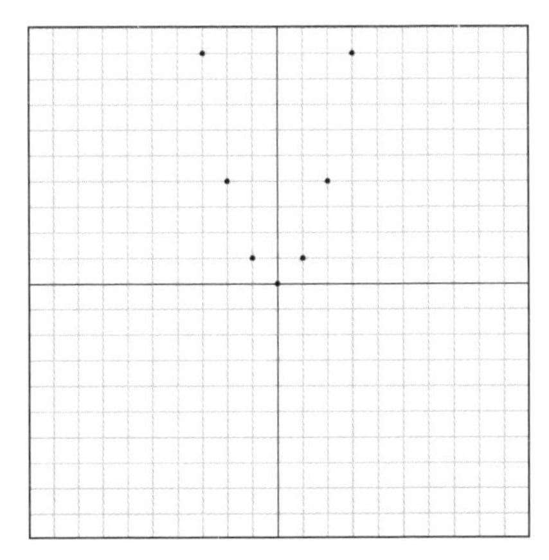

Figure 4-6: A graph of disconnected dots.

Using a different kind of loop, we could draw dots closer together, as in Figure 4-7.

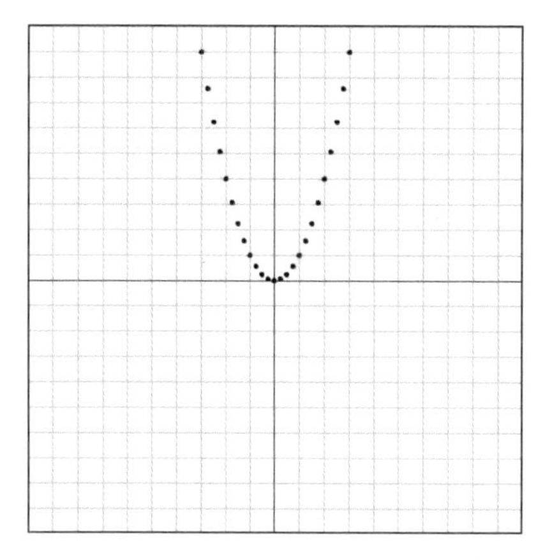

Figure 4-7: The dots are closer together, but it's still not a convincing curve.

The best way to make a connected curve is to draw lines from point to point. If the points are close enough together, they'll look curved. First, we'll create a `graphFunction()` function after `f(x)`.

Connecting the Points

In the `graphFunction()` function, start `x` at `xmin`, like this:

grid.pyde

```
def graphFunction():
    x = xmin
```

To make the graph extend across the whole grid, we'll keep increasing `x` until it's equal to `xmax`. That means we'll keep this loop going "while `x` is less than or equal to `xmax`," as shown here:

```
def graphFunction():
    x = xmin
    while x <= xmax:
```

To draw the curve itself, we'll draw lines from every point to every next point, going up a tenth of a unit at a time. Even if our function produces a curve, you probably won't notice if we're drawing a straight line between two points that are really close together. For example, the distance from (2, f(2)) to (2.1, f(2.1)) is tiny, so overall the output will look curved.

```
def graphFunction():
    x = xmin
    while x <= xmax:
        fill(0)
        line(x*xscl,f(x)*yscl,(x+0.1)*xscl,f(x+0.1)*yscl)
        x += 0.1
```

This code defines a function that draws a graph of `f(x)` by starting at `xmin` and going all the way up to `xmax`. While the x-value is less than or equal to `xmax`, we'll draw a line from (x, f(x)) to ((x + 0.1), f(x + 0.1)). We can't forget to increment `x` by 0.1 at the end of the loop.

Listing 4-11 shows the whole code for *grid.pyde*.

grid.pyde

```
#set the range of x-values
xmin = -10
xmax = 10

#range of y-values
ymin = -10
ymax = 10

#calculate the range
rangex = xmax - xmin
rangey = ymax - ymin

def setup():
    global xscl, yscl
    size(600,600)
    xscl = width / rangex
    yscl = -height / rangey

def draw():
    global xscl, yscl
    background(255) #white
    translate(width/2,height/2)
    grid(xscl,yscl)
    graphFunction()

def f(x):
    return x**2

def graphFunction():
    x = xmin
    while x <= xmax:
        fill(0)
        line(x*xscl,f(x)*yscl,(x+0.1)*xscl,f(x+0.1)*yscl)
        x += 0.1

def grid(xscl, yscl):
    #Draws a grid for graphing
    #cyan lines
    strokeWeight(1)
    stroke(0,255,255)
    for i in range(xmin,xmax+1):
        line(i*xscl,ymin*yscl,i*xscl,ymax*yscl)
    for i in range(ymin,ymax+1):
        line(xmin*xscl,i*yscl,xmax*xscl,i*yscl)
    stroke(0) #black axes
```

```
    line(0,ymin*yscl,0,ymax*yscl)
    line(xmin*xscl,0,xmax*xscl,0)
```

Listing 4-11: Complete code for graphing the parabola

This gets us the curve we're looking for, as shown in Figure 4-8.

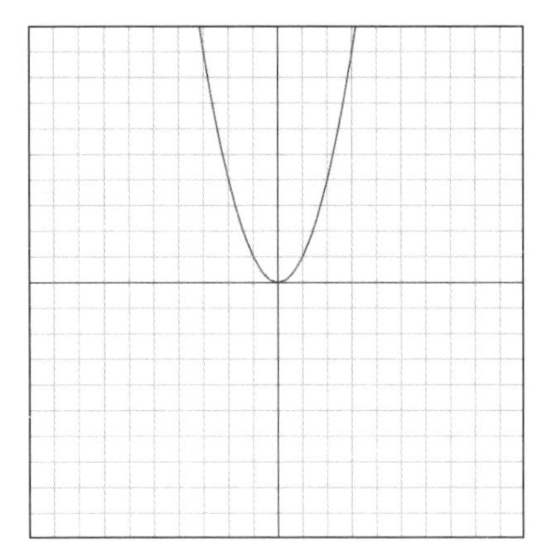

Figure 4-8: A nice continuous graph of a parabola!

Now we can change our function to something more complicated, and the grapher will easily draw it:

grid.pyde

```
def f(x):
    return 6*x**3 + 31*x**2 + 3*x - 10
```

With this simple change, you'll see the output in Figure 4-9, but the function will be in black. If you prefer a red curve, change the `stroke(0)` line in `graphFunction()` to `stroke(255,0,0)`, and you'll get a red function.

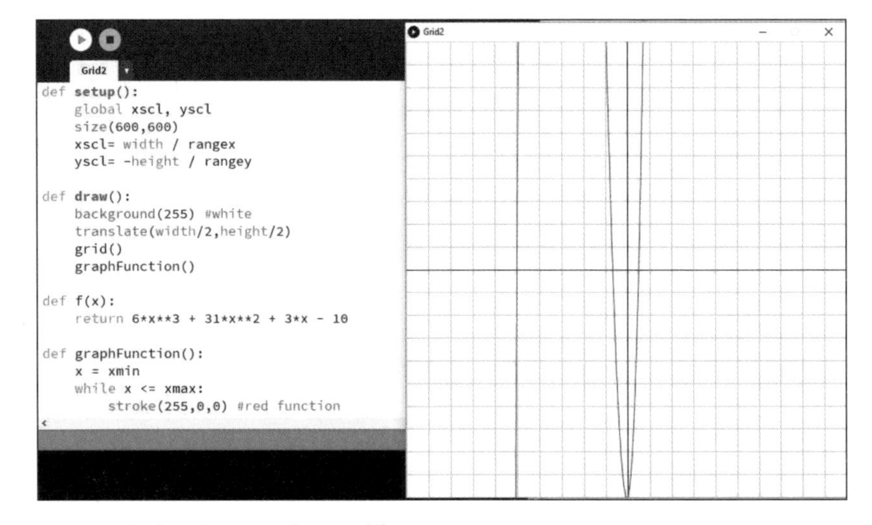

Figure 4-9: Graphing a polynomial function

It's great to be able to simply change one line in the f() function and have the program automatically graph a different function! The solutions (called the *roots*) of the equation are where the graph crosses the x-axis. We can see three places: one where $x = -5$, another where x is between –1 and 0, and a third where x is between 0 and 1.

USING GUESS AND CHECK TO FIND THE ROOTS

We already saw how effective our guess-and-check method was for guessing numbers in Chapter 3. Now we can use it to approximate the roots, or solutions, for the equation $6x^3 + 31x^2 + 3x - 10 = 0$. Let's start with the root between 0 and 1. Is it 0.5 or something else? To test this, we can easily plug 0.5 into the equation. Create a new file in IDLE, name it *guess.py*, and enter the following code:

guess.py

```
def f(x):
    return 6*x**3 + 31*x**2 + 3*x - 10

>>> f(0.5)
0.0
```

As you can see, when x equals 0.5, it makes the function equal 0, so another solution of our equation is $x = 0.5$.

Next, let's try to find the root between –1 and 0. We'll try the average of –1 and 0 to start:

```
>>> f(-0.5)
-4.5
```

At $x = -0.5$, the function is negative, not zero. Looking at the graph, we can tell we guessed too high, so the root must be somewhere between –1 and –0.5. We'll average those endpoints and try again:

```
>>> f(-0.75)
2.65625
```

We get a positive number, so we guessed too low. Therefore, the solution must be between –0.75 and –0.5:

```
>>> f(-0.625)
-1.23046875
```

Still too high. This is getting a bit tedious. Let's see how we might use Python to do these steps for us.

WRITING THE GUESS() FUNCTION

Let's create a function that will find the roots of an equation by averaging the lower and upper values and adjusting its next guesses accordingly. This will work for our current task, where the function is passing through the x-axis from positive to negative. For a function going up, from negative to positive, we'd have to change it around a little. Listing 4-12 shows the complete code for this function.

```
'''The guess method'''
def f(x):
    return 6*x**3 + 31*x**2 + 3*x - 10

def average(a,b):
    return (a + b)/2.0

def guess():
    lower = -1
    upper = 0
 ❶ for i in range(20):
        midpt = average(lower,upper)
        if f(midpt) == 0:
            return midpt
        elif f(midpt) < 0:
            upper = midpt
        else:
            lower = midpt
    return midpt

x = guess()

print(x,f(x))
```

Listing 4-12: The guess method for solving equations

First, we declare the function for the equation we're trying to solve using `f(x)`. Then we create the `average()` function to find the average of two numbers, which we'll be using at every step. Finally, we write a `guess()` function that starts with a lower limit of –1 and an upper limit of 0, since that's where our graph crossed the x-axis.

We then use `for i in range(20):` ❶ to create a loop that cuts the range by half 20 times. Our guess will be the average, or midpoint, of the upper and lower limits. We put that midpoint into `f(x)` and if the output equals 0, we know that's our root. If the output is negative, we know we guessed too high. Then the midpoint will replace our upper limit and we'll take another guess. Otherwise, if we guessed too low, the midpoint will become our lower limit and we'll guess again.

If we haven't returned the solution in 20 guesses, we return the latest midpoint and the function of that midpoint.

When we run this, we should get two values as the output:

```
-0.6666669845581055 9.642708896251406e-06
```

The first output is the x-value, which is very close to –2/3. The second output is what `f(x)` evaluates to when we plug in –2/3 as the x-value. The `e-06` at the end is scientific notation, which means you take 9.64 and move the decimal place to the left six places. So `f(x)` evaluates to 0.00000964, which is very close to zero. To go through this guess-and-check program and get this solution, or rather an approximation accurate to within a millionth of the actual solution, to pop up in less than a second is still surprising and wonderful to me! Can you see the power in exploring math problems using free software like Python and Processing?

If we increase the number of iterations from 20 to 40, we get a number even closer to 0:

```
-0.6666666666669698 9.196199357575097e-12
```

Let's check `f(-0.6666666666669698)`, or `f(-2/3)`:

```
>>> f(-2/3)
0.0
```

This checks out, so the three solutions to $6x^3 + 31x^2 + 3x - 10 = 0$ are $x = -5$, $-2/3$, and $1/2$.

EXERCISE 4-3: FINDING MORE ROOTS

Use the graphing tool you just created to find the roots of $2x^2 + 7x - 15 = 0$. Remember, the roots are where the graph crosses the x-axis, or where the function equals 0. Check your answers using your quad() function.

SUMMARY

Math class used to be all about taking years to learn how to solve equations of higher and higher degree. In this chapter you learned that this isn't so hard to do programmatically using our guess-and-check method. You also wrote programs that solve equations in other ways, like using the quadratic formula and graphing. In fact, you learned that all we have to do to solve an equation, no matter how complicated, is to graph it and approximate where it crosses the x-axis. By iterating and halving the range of values that work, we can get as accurate as we want.

In programming, we use algebra to create variables to represent values that will change, like the size or coordinates of an object. The user can then change the value of a variable in one place, and the program will automatically change the value of that variable everywhere in the program. The user can also change these variables using a loop or declare the value in a function call. In future chapters we'll model real-life situations where we need to use variables to represent parameters and constraints on the model, like energy content and force of gravity. Using variables lets us change values easily, to vary different aspects of the model.

In the next chapter you'll use Processing to create interactive graphics, like rotating triangles and colorful grids!

TRANSFORMING SHAPES WITH GEOMETRY

In the teahouse one day Nasrudin announced he was selling his house. When the other patrons asked him to describe it, he brought out a brick. "It's just a collection of these."
—Idries Shah

In geometry class, everything you learn about involves dimensions in space using shapes. You typically start by examining one-dimensional lines and two-dimensional circles, squares, or triangles, then move on to three-dimensional objects like spheres and cubes. These days, creating geometric shapes is easy with technology and free software, though manipulating and changing the shapes you create can be more of a challenge.

In this chapter, you'll learn how to manipulate and transform geometric shapes using the Processing graphics package. You'll start with basic shapes like circles and triangles, which will allow you to work with complicated shapes like fractals and cellular automata in later chapters. You will also learn how to break down some complicated-looking designs into simple components.

DRAWING A CIRCLE

Let's start with a simple one-dimensional circle. Open a new sketch in Processing and save it as *geometry.pyde*. Then enter the code in Listing 5-1 to create a circle on the screen.

geometry.pyde

```
def setup():
    size(600,600)

def draw():
    ellipse(200,100,20,20)
```

Listing 5-1: Drawing a circle

Before we draw the shape, we first define the size of our sketchbook, known as the *coordinate plane*. In this example, we use the `size()` function to say that our grid will be 600 pixels wide and 600 pixels tall.

With our coordinate plane set up, we then use the drawing function `ellipse()` to create our circle on this plane. The first two parameters, `200` and `100`, show where the center of the circle is located. Here, `200` is the x-coordinate and the second number, `100`, is the y-coordinate of this circle's center, which places it at `(200,100)` on the plane.

The last two parameters determine the width and height of the shape in pixels. In the example, the shape is 20 pixels wide and 20 pixels tall. Because the two parameters are the same, it means that the points on the circumference are equidistant from the center, forming a perfectly round circle.

Click the **Run** button (it looks like a play symbol), and a new window with a small circle should open, like in Figure 5-1.

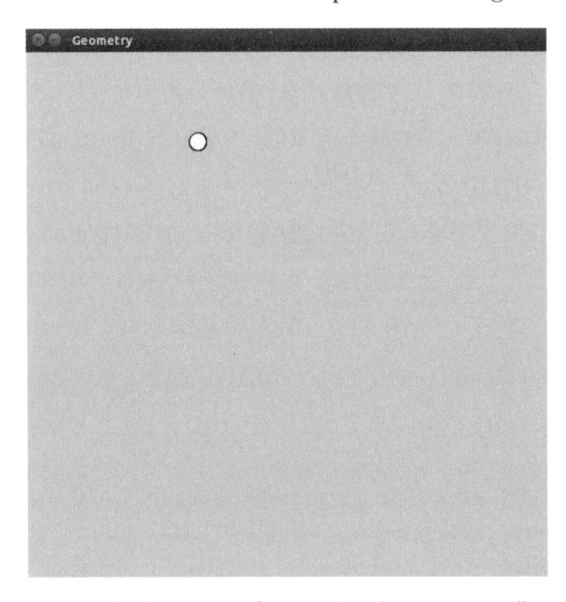

Figure 5-1: The output of Listing 5-1 showing a small circle

Processing has a number of functions you can use to draw shapes. Check out the full list at *https://processing.org/reference/* to explore other shape functions.

Now that you know how to draw a circle in Processing, you're almost ready to use these simple shapes to create dynamic, interactive graphics. In order to do that, you'll first need to learn about location and transformations. Let's start with location.

SPECIFYING LOCATION USING COORDINATES

In Listing 5-1, we used the first two parameters of the `ellipse()` function to specify our circle's location on the grid. Likewise, each shape we create using Processing needs a location that we specify with the coordinate system, where each point on the graph is represented by two numbers: (x,y). In traditional math graphs, the origin (where x=0 and y=0) is at the center of the graph, as shown in Figure 5-2.

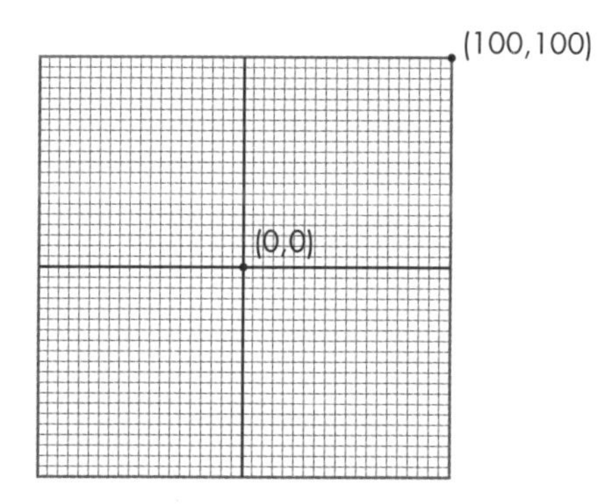

Figure 5-2: A traditional coordinate system with the origin in the center

In computer graphics, however, the coordinate system is a little different. Its origin is in the top-left corner of the screen so that x and y increase as you move right and down, respectively, as you can see in Figure 5-3.

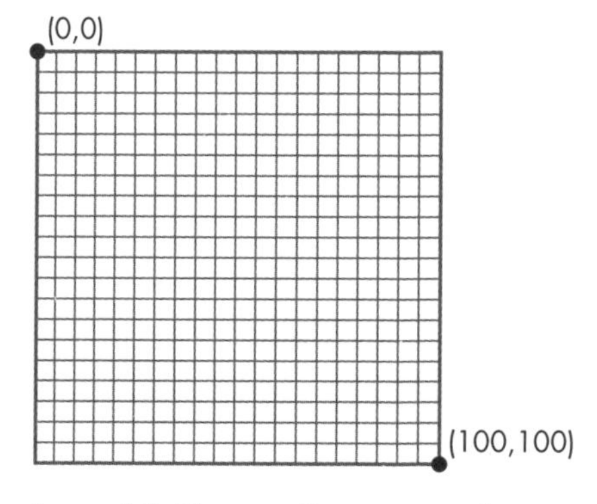

Figure 5-3: The coordinate system for computer graphics, with the origin in the top-left corner

Each coordinate on this plane represents a pixel on the screen. As you can see, this means you don't have to deal with negative coordinates. We'll use functions to transform and translate increasingly complex shapes around this coordinate system.

Drawing a single circle was fairly easy, but drawing multiple shapes can get complicated pretty quickly. For example, imagine drawing a design like the one shown in Figure 5-4.

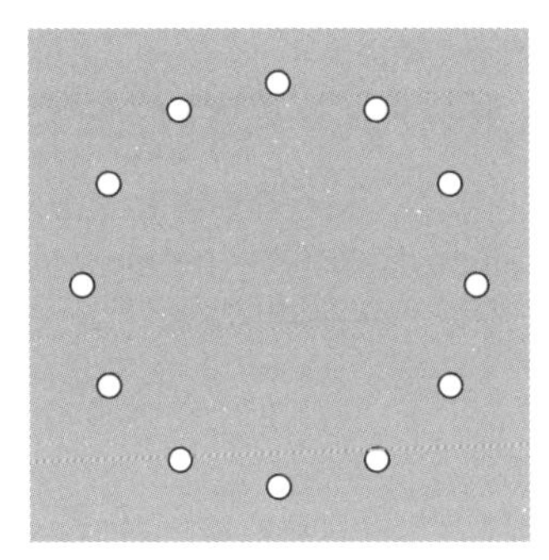

Figure 5-4: A circle made of circles

Specifying the size and location of each individual circle and spacing them out perfectly evenly would involve entering many lines of similar code. Fortunately, you don't really need to know the absolute x- and y-coordinates of each circle to do this. With Processing, you can easily place objects wherever you want on the grid.

Let's see how you can do this using a simple example to start.

TRANSFORMATION FUNCTIONS

You might remember doing transformations with pencil and paper in geometry class, which you performed on a collection of points to laboriously move a shape around. It's much more fun when you let a computer do the transforming. In fact, there wouldn't be any computer graphics worth looking at without transformations! Geometric transformations like translation and rotation let you change where and how your objects appear without altering the objects themselves. For example, you can use transformations to move a triangle to a different location or spin it around without changing its shape. Processing has a number of built-in transformation functions that make it easy to translate and rotate objects.

TRANSLATING OBJECTS WITH TRANSLATE()

To *translate* means to move a shape on a grid so that all points of the shape move in the same direction and the same distance. In other words, translations let you move a shape on a grid without changing the shape itself and without tilting it in the slightest.

Translating an object in math class involves manually changing the coordinates of all the points in the object. But in Processing, you translate an object by moving the *grid* itself, while the object's coordinates stay the same! For an example of this, let's put a rectangle on the screen. Revise your existing code in *geometry.pyde* with the code in Listing 5-2.

geometry.pyde

```
def setup():
    size(600,600)

def draw():
    rect(20,40,50,30)
```

Listing 5-2: Drawing a rectangle to translate

Here, we use the `rect()` function to draw the rectangle. The first two parameters are the x- and y-coordinates telling Processing where the top-left corner of the rectangle should be. The third and fourth parameters indicate its width and its height, respectively.

Run this code, and you should see the rectangle shown in Figure 5-5.

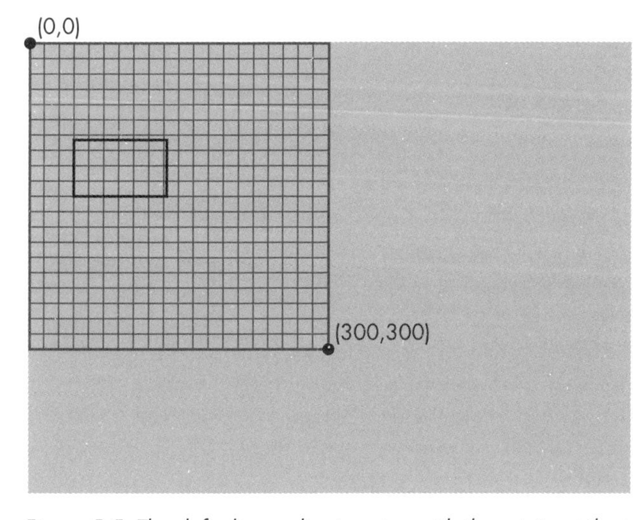

Figure 5-5: The default coordinate setup with the origin at the top left

NOTE *In these examples, I'm showing the grid for reference, but you won't see it on your screen.*

Now let's tell Processing to translate the rectangle using the code in Listing 5-3. Notice that we don't change the coordinates of the rectangle.

geometry.pyde

```
def setup():
    size(600,600)

def draw():
    translate(50,80);
    rect(50,100,100,60)
```

Listing 5-3: Translating the rectangle

Here, we use translate() to move the rectangle. We provide two parameters: the first tells Processing how far to move the grid in the horizontal (x) direction, and the second parameter is for how far to move the grid vertically, in the y-direction. So translate(50,80) should move the entire grid 50 pixels to the right and 80 pixels down, as shown in Figure 5-6.

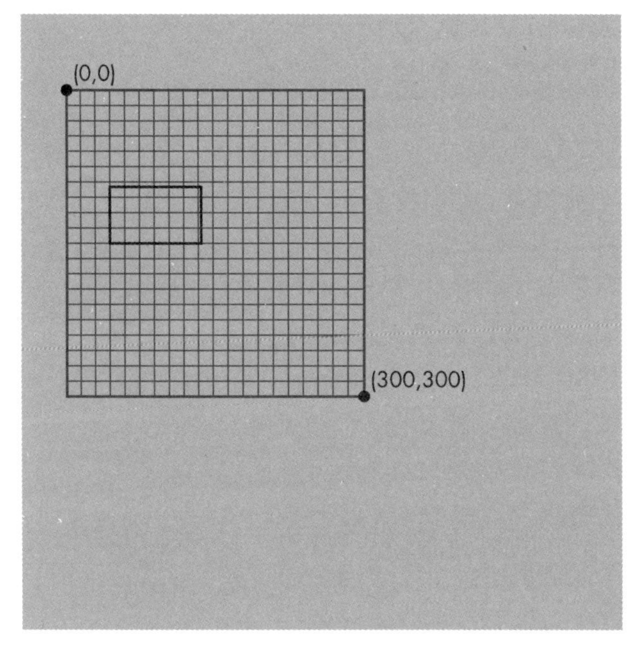

Figure 5-6: Translating a rectangle by moving the grid 50 pixels to the right and 80 pixels down

Very often it's useful (and easier!) to have the origin (0,0) in the center of the canvas. You can use translate() to easily move the origin to the center of your grid. You can also use it to change the width and height of your canvas if you want it bigger or smaller. Let's explore Processing's built-in width and height variables, which let you update the size of your canvas without having to change the numbers manually. To see this in action, update the existing code in Listing 5-3 so it looks like Listing 5-4.

geometry.pyde

```
def setup():
    size(600,600)

def draw():
    translate(width/2, height/2)
    rect(50,100,100,60)
```

Listing 5-4: Using the width and height variables to translate the rectangle

Whatever numbers you put in the size declaration in the setup() function will become the "width" and "height" of the canvas. In this case, because I used size(600,600), they're both 600 pixels. When we change the translate() line to translate(width/2, height/2) using variables instead of specific numbers, we tell Processing to move the location (0,0) to the center of the display window, no matter what the size is. This means that if you change the size of the window, Processing will automatically update `width` and `height`, and you won't have to go through all your code and change the numbers manually.

Run the updated code, and you should see something like Figure 5-7.

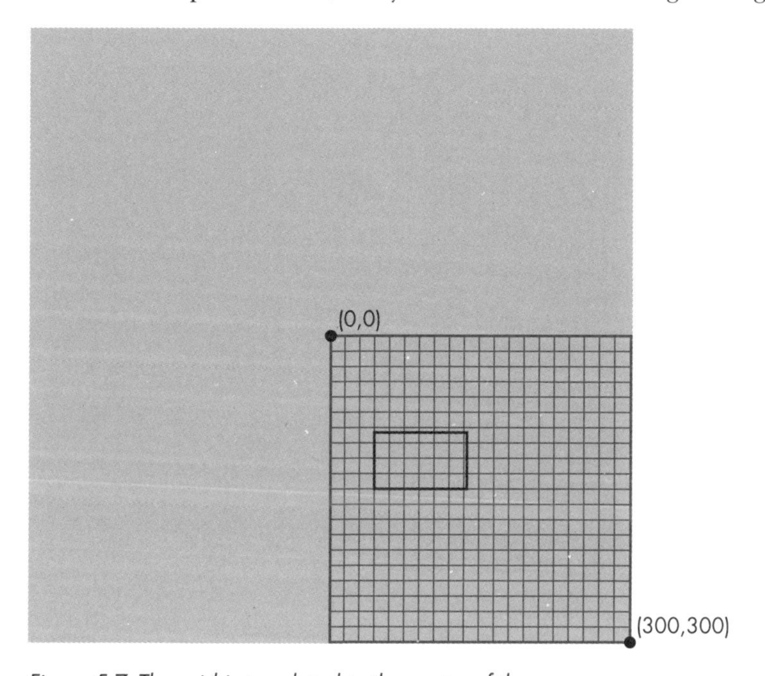

Figure 5-7: The grid is translated to the center of the screen.

Notice that the origin is still labeled as (0,0), which shows that we haven't actually moved the origin point but rather the entire coordinate plane itself so that the origin point falls in the middle of our canvas.

ROTATING OBJECTS WITH ROTATE()

In geometry, *rotation* is a kind of transformation that turns an object around a center point, as if it's turning on an axis. The `rotate()` function in Processing rotates the grid around the origin (0,0). It takes a single number as its argument to specify the angle at which you want to rotate the grid around the point (0,0). The units for the rotation angle are radians, which you learn about in precalculus class. Instead of using 360 degrees to do a full rotation, we can use 2π (around 6.28) radians. If you think in degrees, like I do, you can use the `radians()` function to easily convert your degrees to radians so you don't have to do the math yourself.

To see how the `rotate()` function works, enter the code shown in Figure 5-8 into your existing sketch by replacing the `translate()` code inside the `draw()` function with each of these examples, and then run them. Figure 5-8 shows the results.

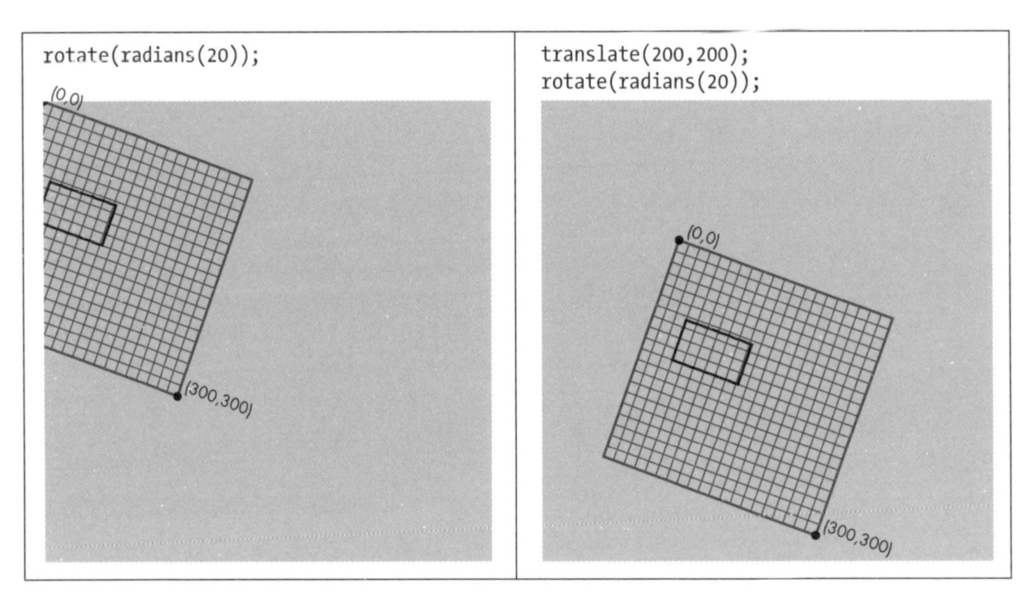

Figure 5-8: The grid always rotates around (0,0)

On the left side of Figure 5-8, the grid is rotated 20 degrees around (0,0), which is at the top-left corner of the screen. In the example on the right, the origin is first translated 200 units to the right and 200 units down and *then* the grid is rotated.

The `rotate()` function makes it easy to draw a circle of objects like the one in Figure 5-4 using the following steps:

1. Translate to where you want the center of the circle to be.
2. Rotate the grid and put the objects along the circumference of the circle.

Now that you know how to use transformation functions to manipulate the location of different objects on your canvas, let's actually re-create Figure 5-4 in Processing.

DRAWING A CIRCLE OF CIRCLES

To create the circles arranged in a circle in Figure 5-4, we'll use a `for i in range()` loop to repeat the circles and make sure the circles are evenly spaced. First, let's think about how many degrees should be between the circles to make a full circle, remembering that a circle is 360 degrees.

Enter the code shown in Listing 5-5 to create this design.

geometry.pyde

```
def setup():
    size(600,600)

def draw():
    translate(width/2,height/2)
    for i in range(12):
```

```
        ellipse(200,0,50,50)
        rotate(radians(360/12))
```

Listing 5-5: Drawing a circular design

Note that the `translate(width/2,height/2)` function inside the `draw()` function translates the grid to the center of the screen. Then, we start a `for` loop to create an ellipse at a point on the grid, starting at (200,0), as you can see from the first two parameters of the function. Then we set the size of each small circle by setting both the `width` and `height` of the ellipse to `50`. Finally, we rotate the grid by 360/12, or 30 degrees, before creating the next ellipse. Note that we use `radians()` to convert 30 degrees into radians inside the `rotate()` function. This means that each circle will be 30 degrees away from the next one.

When you run this, you should see what's shown in Figure 5-9.

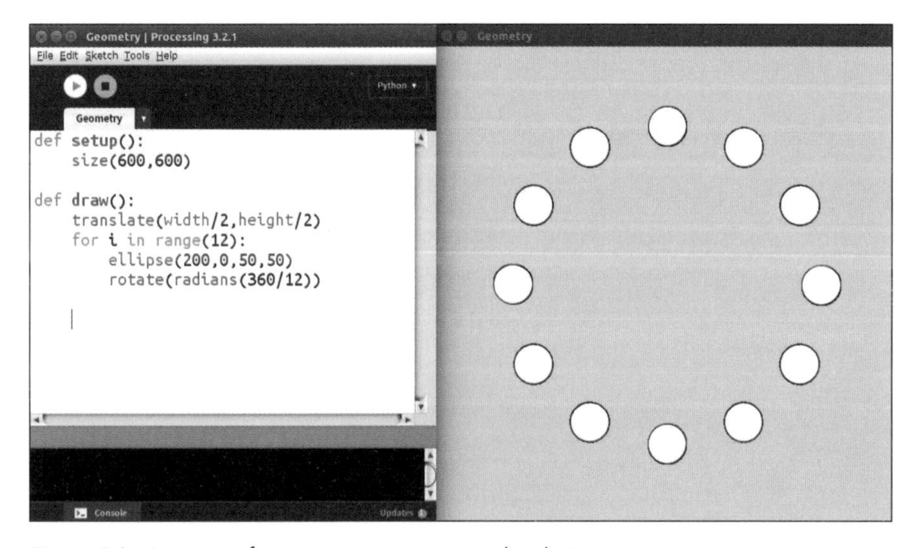

Figure 5-9: Using transformation to create a circular design

We have successfully arranged a bunch of circles into a circular shape!

DRAWING A CIRCLE OF SQUARES

Modify the program you wrote in Listing 5-5 and change the circles into squares. To do this, just change `ellipse` in the existing code to `rect` to make the circles into squares, as shown here:

geometry.pyde

```
def setup():
    size(600,600)

def draw():
    translate(width/2,height/2)
    for i in range(12):
        rect(200,0,50,50)
        rotate(radians(360/12))
```

That was easy!

ANIMATING OBJECTS

Processing is great for animating your objects to create dynamic graphics. For your first animation, you'll use the `rotate()` function. Normally, `rotate` happens instantly, so you don't get to see the action take place—only the result of the rotation. But this time, we'll use a time variable `t`, which allows us to see the rotation unfold in real time!

CREATING THE T VARIABLE

Let's use our circle of squares to write an animated program. To start, create the `t` variable and initialize it to 0 by adding **`t = 0`** before the `setup()` function. Then insert the code in Listing 5-6 before the `for` loop.

geometry.pyde

```
t = 0

def setup():
    size(600,600)

def draw():
    translate(width/2,height/2)
    rotate(radians(t))
    for i in range(12):
        rect(200,0,50,50)
        rotate(radians(360/12))
```

Listing 5-6: Adding the t variable

However, if you try to run this code, you'll get the following error message:

```
UnboundLocalError: local variable 't' referenced before assignment
```

This is because Python doesn't know whether we're creating a new local variable named `t` *inside* the function that doesn't have anything to do with the global variable `t` *outside* the function, or just calling the global variable. Because we want to use the global variable, add **`global t`** at the beginning of the `draw()` function so the program knows which one we're referring to.

Enter the complete code shown here:

geometry.pyde

```
t = 0

def setup():
    size(600,600)

def draw():
    global t
    #set background white
    background(255)
```

```
    translate(width/2,height/2)
    rotate(radians(t))
    for i in range(12):
        rect(200,0,50,50)
        rotate(radians(360/12))
    t += 1
```

This code starts t at 0, rotates the grid that number of degrees, increments t by 1, and then repeats. Run it, and you should see the squares start to rotate in a circular pattern, as in Figure 5-10.

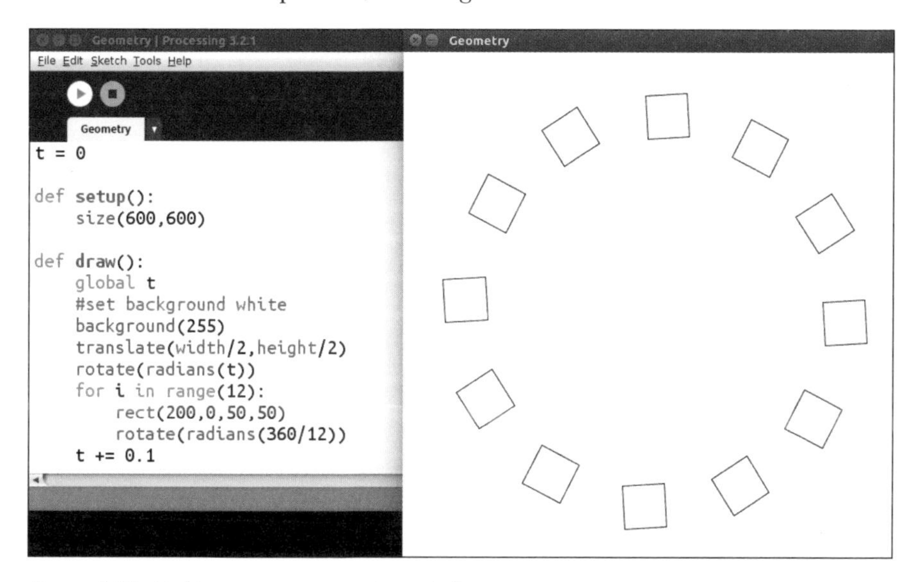

Figure 5-10: Making squares rotate in a circle

Pretty cool! Now let's try rotating each individual square.

ROTATING THE INDIVIDUAL SQUARES

Because rotating is done around (0,0) in Processing, inside the loop we first have to translate to where each square needs to be, then rotate, and finally draw the square. Change the loop in your code to look like Listing 5-7.

geometry.pyde

```
    for i in range(12):
        translate(200,0)
        rotate(radians(t))
        rect(0,0,50,50)
        rotate(radians(360/12))
```

Listing 5-7: Rotating each square

This translates the grid to where we want to place the square, rotates the grid so the square rotates, and then draws the square using the rect() function.

SAVING ORIENTATION WITH PUSHMATRIX() AND POPMATRIX()

When you run Listing 5-7, you should see that it creates some strange behavior. The squares don't rotate around the center, but keep moving around the screen instead, as shown in Figure 5-11.

Figure 5-11: The squares are flying all over!

This is due to changing the center and changing the orientation of the grid so much. After translating to the location of the square, we need to rotate back to the center of the circle before translating to the next square. We could use another `translate()` function to undo the first one, but we might have to undo more transformations, and that could get confusing. Fortunately, there's an easier way.

Processing has two built-in functions that save the orientation of the grid at a certain point and then return to that orientation: `pushMatrix()` and `popMatrix()`. In this case, we want to save the orientation when we're in the center of the screen. To do this, revise the loop to look like Listing 5-8.

geometry.pyde

```
for i in range(12):
    pushMatrix() #save this orientation
    translate(200,0)
    rotate(radians(t))
    rect(0,0,50,50)
    popMatrix() #return to the saved orientation
    rotate(radians(360/12))
```

Listing 5-8: Using `pushMatrix()` and `popMatrix()`

The `pushMatrix()` function saves the position of the coordinate system at the center of the circle of squares. Then we translate to the location of the square, rotate the grid so the square will spin, and then draw the square. Then we use popMatrix() to return instantly to the center of the circle of squares and repeat for all 12 squares.

ROTATING AROUND THE CENTER

The preceding code should work perfectly, but the rotation may look strange; that's because Processing by default locates a rectangle at its top-left corner and rotates it about its top-left corner. This makes the squares look like they're veering off the path of the larger circle. If you want your squares to rotate around their centers, add this line to your setup() function:

```
rectMode(CENTER)
```

Note that the all-uppercase CENTER in rectMode() matters. (You can also experiment with other types of rectMode(), like CORNER, CORNERS, and RADIUS.) Adding rectMode(CENTER) should make each square rotate around its center. If you want the squares to spin more quickly, change the rotate() line to increase the time in t, like so:

```
rotate(radians(5*t))
```

Here, 5 is the frequency of the rotation. This means the program multiplies the value of t by 5 and rotates by the product. Therefore, the square will rotate five times as far as before. Change it to see what happens! Comment out the rotate() line outside the loop (by adding a hashtag at the beginning) to make the squares rotate in place, as shown in Listing 5-9.

```
translate(width/2,height/2)
#rotate(radians(t))
for i in range(12):
    rect(200,0,50,50)
```

Listing 5-9: Commenting out a line instead of deleting it

Being able to use transformations like translate() and rotate() to create dynamic graphics is a very powerful technique, but it can produce unexpected results if you do things in the wrong order!

CREATING AN INTERACTIVE RAINBOW GRID

Now that you've learned how to create designs using loops and to rotate them in different ways, we'll create something pretty awesome: a grid of squares whose rainbow colors follow your mouse cursor! The first step is to make a grid.

DRAWING A GRID OF OBJECTS

Many tasks involved in math and in creating games like Minesweeper require a grid. Grids are necessary for some of the models and all the cellular automata we'll create in later chapters, so it's worth learning how to write code for making a grid that we can reuse. To begin with, we'll make a 12 × 12 grid of squares, evenly sized and spaced. Making a grid this size may seem like a time-consuming task, but in fact it's easy to do using a loop.

Open a new Processing sketch and save as *colorGrid.pyde*. Too bad we used the name "grid" previously. We'll make a 20 × 20 grid of squares on a white background. The squares need to be `rect`, and we need to use a `for` loop within a `for` loop to make sure they are all the same size and spaced equally. Also, we need our 25 × 25 pixel squares to be drawn every 30 pixels, using this line:

```
rect(30*x,30*y,25,25)
```

As the `x` and `y` variables go up by 1, squares are drawn at 50-pixel intervals in two dimensions. We'll start off, as usual, by writing our `setup()` and `draw()` functions, as in the previous sketch (see Listing 5-10).

colorGrid.pyde

```
def setup():
    size(600,600)

def draw():
    #set background white
    background(255)
```

Listing 5-10: The standard structure for a Processing sketch: setup() and draw()

This sets the size of the window at 600 by 600 pixels, and sets the background color to white. Next we'll create a nested loop, where two variables will both go from 0 to 19, for a total of 20 numbers, since we want 20 rows of 20 squares. Listing 5-11 shows the code that creates the grid.

```
def setup():
    size(600,600)

def draw():
    #set background white
    background(255)
    for x in range(20):
        for y in range(20):
            rect(30*x,30*y,25,25)
```

Listing 5-11: The code for a grid

This should create a 20 × 20 grid of squares, as you can see in Figure 5-12. Time to add some colors to our grid.

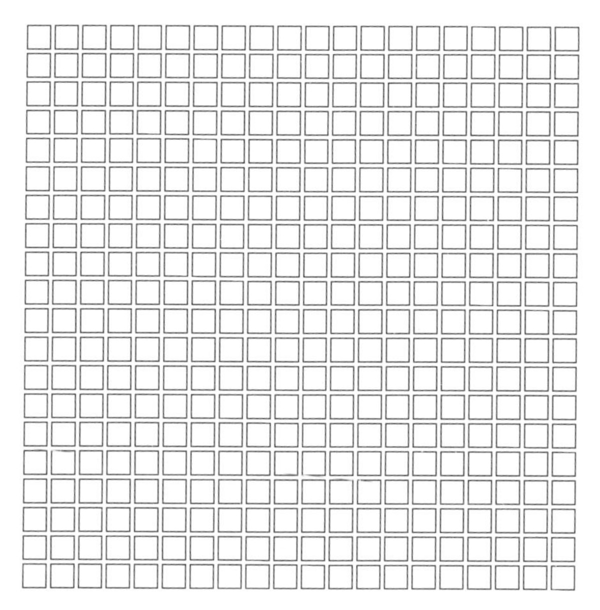

Figure 5-12: A 20 × 20 grid!

ADDING THE RAINBOW COLOR TO OBJECTS

Processing's `colorMode()` function helps us add some cool color to our sketches! It's used to switch between the RGB and HSB modes. Recall that RGB uses three numbers indicating amounts of red, green, and blue. In HSB, the three numbers represent levels of hue, saturation, and brightness. The only one we need to change here is the first number, which represents the hue. The other two numbers can be the maximum value, 255. Figure 5-13 shows how to make rainbow colors by changing only the first value, the hue. Here, the 10 squares have the hue values shown in the figure, with 255 for saturation and 255 for brightness.

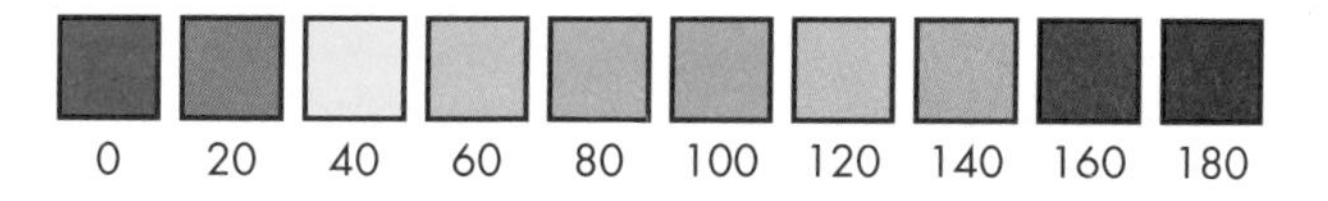

Figure 5-13: The colors of the rainbow using HSB mode and changing the hue value

Since we're locating the rectangles at `(30*x,30*y)` in Listing 5-11, we'll create a variable that measures the distance of the mouse from that location:

```
d = dist(30*x,30*y,mouseX,mouseY)
```

Processing has a `dist()` function that finds the distance between two points, and in this case it's the distance between the square and the mouse. It saves the distance to a variable called `d`, and we'll link the hue to that variable. Listing 5-12 shows the changes to the code.

colorGrid.pyde

```
def setup():
    size(600,600)
    rectMode(CENTER)
  ❶ colorMode(HSB)

def draw():
    #set background black
  ❷ background(0)
    translate(20,20)
    for x in range(30):
        for y in range(30):
          ❸ d = dist(30*x,30*y,mouseX,mouseY)
            fill(0.5*d,255,255)
            rect(30*x,30*y,25,25)
```

Listing 5-12: Using the `dist()` function

We insert the `colorMode()` function and pass `HSB` to it ❶. In the `draw()` function, we set the background to black first ❷. Then we calculate the distance from the mouse to the square, which is at `(30*x,30*y)` ❸. In the next line, we set the fill color using HSB numbers. The hue value is half the distance, while the saturation and brightness numbers are both 255, the maximum.

The hue is the only thing we change: we update the hue according to the distance the rectangle is from the mouse. We do this with the `dist()` function, which takes four arguments: the x- and y-coordinates of two points. It returns the distance between the points.

Run this code and you should see a very colorful design that changes colors according to the mouse's location, as shown in Figure 5-14.

Now that you've learned how to add colors to your objects, let's explore how we can create more complicated shapes.

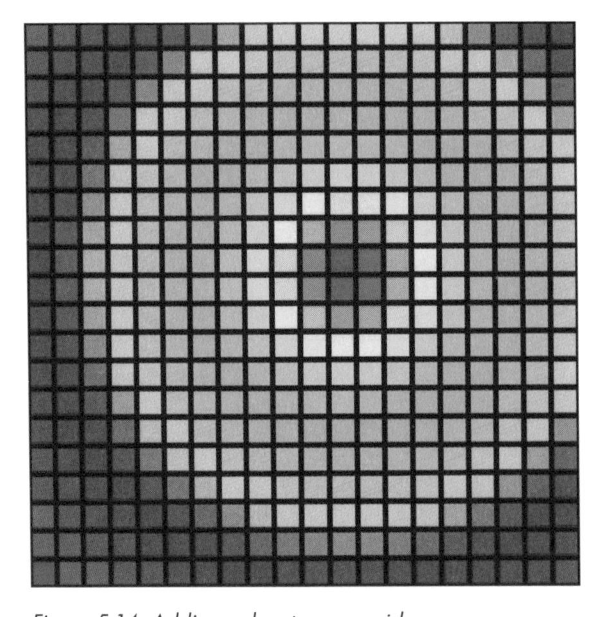

Figure 5-14: Adding colors to your grid

DRAWING COMPLEX PATTERNS USING TRIANGLES

In this section, we create more complicated, Spirograph-style patterns using triangles. For example, take a look at the sketch made up of rotating triangles in Figure 5-15, created by the University of Oslo's Roger Antonsen.

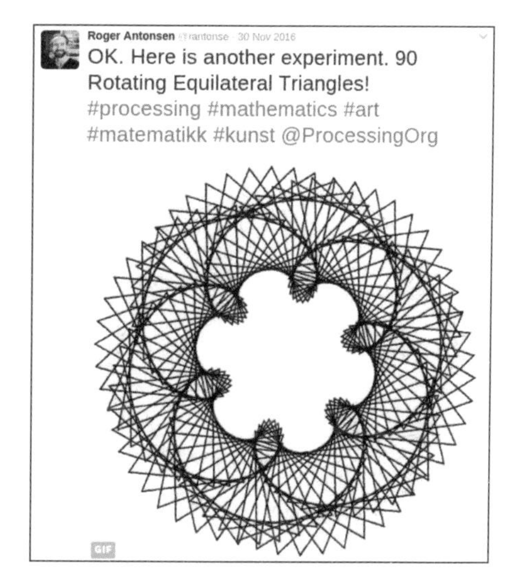

Figure 5-15: Sketch of 90 rotating equilateral triangles by Roger Antonsen. See it in motion at https://rantonse.no/en/art/2016-11-30.

The original design moves, but in this book you'll have to imagine all the triangles rotating. This sketch blew me away! Although this design looks very complicated, it's not that difficult to make. Remember Nasrudin's joke about the brick from the beginning of the chapter? Like Nasrudin's house, this complicated design is just a collection of identical shapes. But what shape? Antonsen gave us a helpful clue to creating this design when he named the sketch "90 Rotating Equilateral Triangles." It tells us that all we have to do is figure out how to draw an equilateral triangle, rotate it, and then repeat that for a total of 90 triangles. Let's first discuss how to draw an equilateral triangle using the `triangle()` function. To start, open a new Processing sketch and name it *triangles.pyde*. The code in Listing 5-13 shows one way to create a rotating triangle but not an equilateral one.

triangles.pyde

```
def setup():
    size(600,600)
    rectMode(CENTER)

t = 0

def draw():
    global t
    translate(width/2,height/2)
    rotate(radians(t))
    triangle(0,0,100,100,200,-200)
    t += 0.5
```

Listing 5-13: Drawing a rotating triangle, but not the right kind

Listing 5-13 uses the lessons you learned previously: it creates a `t` variable (for time), translates to where we want the triangle to be, rotates the grid, and then draws the triangle. Finally, it increments `t`. When you run this code, you should see something like Figure 5-16.

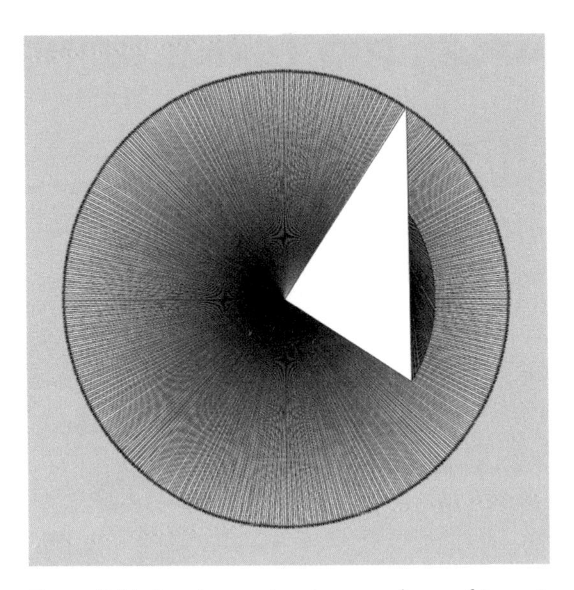

Figure 5-16: Rotating a triangle around one of its vertices

As you can see in Figure 5-16, the triangle rotates around one of its *vertices*, or points, and thus creates a circle with the outer point. You'll also notice that this is a right triangle (a triangle containing a 90-degree angle), not an equilateral one.

To re-create Antonsen's sketch, we need to draw an equilateral triangle, which is a triangle with equal sides. We also need to find the center of the equilateral triangle to be able to rotate it about its center. To do this, we need to find the location of the three vertices of the triangle. Let's discuss how to draw an equilateral triangle by locating it at its center and specifying the location of its vertices.

A 30-60-90 TRIANGLE

To find the location of the three vertices of our equilateral triangle, we'll review a particular type of triangle you've likely seen in geometry class: the *30-60-90 triangle*, which is a special *right triangle*. First, we need an equilateral triangle, as shown in Figure 5-17.

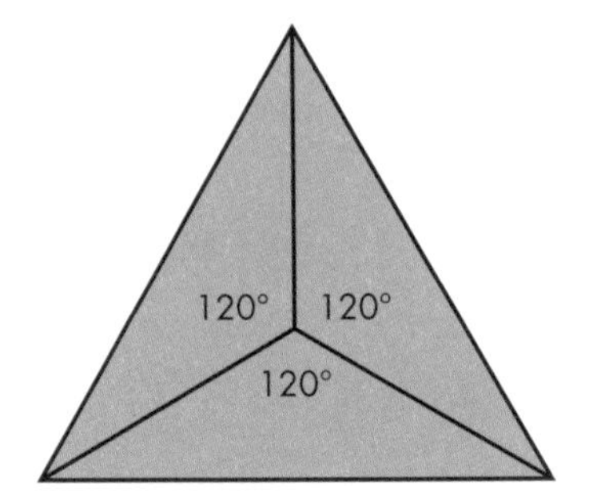

Figure 5-17: An equilateral triangle divided into three equal parts

This equilateral triangle is made up of three equal parts. The point in the middle is the center of the triangle, with the three dissecting lines meeting at 120 degree angles. To draw a triangle in Processing, we give the `triangle()` function six numbers: the x- and y-coordinates of all three vertices. To find the coordinates of the vertices of the equilateral triangle shown in Figure 5-17, let's cut the bottom triangle in half, as shown in Figure 5-18.

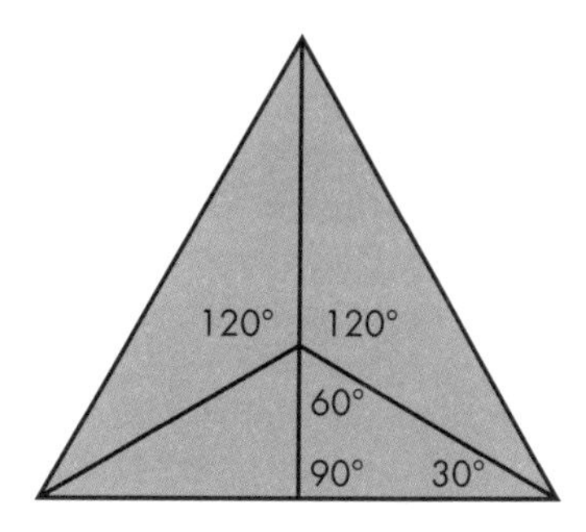

Figure 5-18: Dividing up the equilateral triangle into special triangles

Dividing the bottom triangle in half creates two right triangles, which are classic 30-60-90 triangles. As you might recall, the ratio between the sides of a 30-60-90 triangle can be expressed as shown in Figure 5-19.

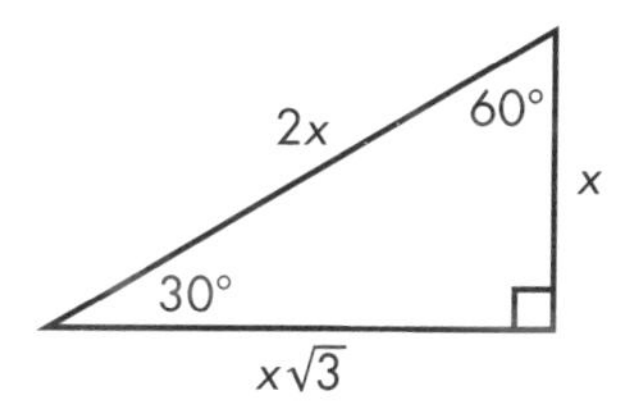

Figure 5-19: The ratios of the sides in a 30-60-90 triangle, from the legend on an SAT test

If we call the length of the smaller leg x, the hypotenuse is twice that length, or $2x$, and the longer leg is x times the square root of 3, or approximately $1.732x$. We're going to be creating our function using the length from the center of the big equilateral triangle in Figure 5-18 to one of its vertices, which happens to be the hypotenuse of the 30-60-90 triangle. That means we can measure everything in terms of that length. For example, if we call the hypotenuse `length`, then the smaller leg will be half that length, or `length/2`. Finally, the longer leg will be `length` divided by 2 times the square root of 3. Figure 5-20 zooms in on the 30-60-90 triangle.

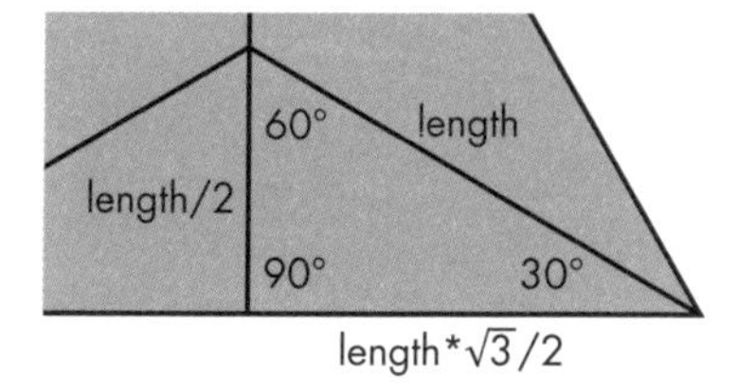

Figure 5-20: The 30-60-90 triangle up close and personal

As you can see, a 30-60-90 triangle has internal angles of 30, 60, and 90 degrees, and the lengths of the sides are in known proportions. You may be familiar with this from the Pythagorean Theorem, which will come up again shortly.

We'll call the distance from the center of the larger equilateral triangle to its vertex the "length," which is also the *hypotenuse* of the 30-60-90 triangle. You'll need to know the ratios between the lengths of the sides of this special triangle in order to find the three vertices of the equilateral triangle with respect to the center—you can draw it (the big equilateral triangle we're trying to draw) by specifying where each point of the triangle should be.

The shorter leg of the right triangle opposite the 30 degree angle is always half the hypotenuse, and the longer leg is the measure of the shorter leg times the square root of 3. So if we use the center point for drawing the big equilateral triangle, the coordinates of the three vertices would be as shown in Figure 5-21.

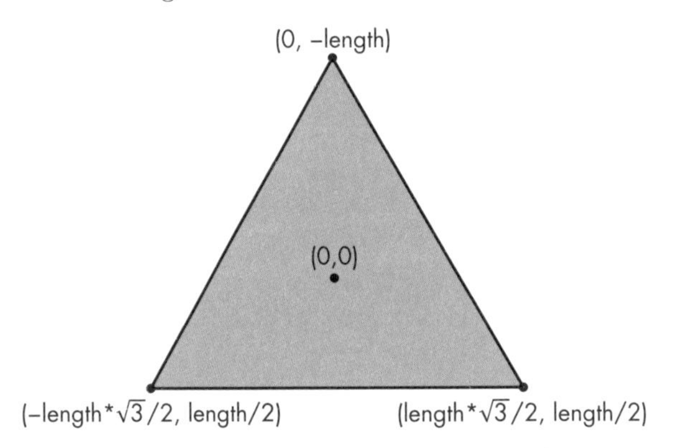

Figure 5-21: The vertices of the equilateral triangle

As you can see, because this triangle is made up of 30-60-90 triangles on all sides, we can use the special relation between them to figure out how far each vertex of the equilateral triangle should be from the origin.

DRAWING AN EQUILATERAL TRIANGLE

Now we can use the vertices we derived from the 30-60-90 triangle to create an equilateral triangle, using the code in Listing 5-14.

triangles.pyde

```
def setup():
    size(600,600)
    rectMode(CENTER)

t = 0

def draw():
    global t
    translate(width/2,height/2)
    rotate(radians(t))
    tri(200) #draw the equilateral triangle
    t += 0.5

❶ def tri(length):
```

```
    '''Draws an equilateral triangle
    around center of triangle'''
❷   triangle(0,-length,
             -length*sqrt(3)/2, length/2,
             length*sqrt(3)/2, length/2)
```

Listing 5-14: The complete code for making a rotating equilateral triangle

First, we write the `tri()` function to take the variable `length` ❶, which is the hypotenuse of the special 30-60-90 triangles we cut the equilateral triangle into. We then make a triangle using the three vertices we found. Inside the call to the `triangle()` function ❷, we specify the location of each of the three vertices of the triangle: `(0,-length)`, `(-length*sqrt(3)/2, length/2)`, and `(length*sqrt(3)/2, length/2)`.

When you run the code, you should see something like Figure 5-22.

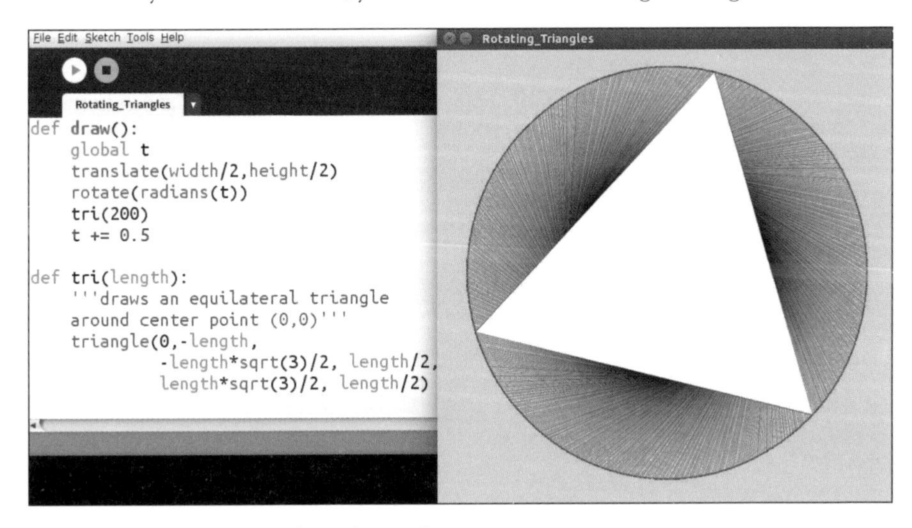

Figure 5-22: A rotating equilateral triangle!

Now we can cover up all the triangles created during rotation by adding this line to the beginning of the `draw()` function:

```
background(255) #white
```

This should erase all the rotating triangles except for one, so we just have a single equilateral triangle on the screen. All we have to do is put 90 of them in a circle, just like we did earlier in this chapter, using the `rotate()` function.

EXERCISE 5-1: SPIN CYCLE

Create a circle of equilateral triangles in a Processing sketch and rotate them using the `rotate()` function.

DRAWING MULTIPLE ROTATING TRIANGLES

Now that you've learned how to rotate a single equilateral triangle, we need to figure out how to arrange multiple equilateral triangles into a circle. This is similar to what you created while rotating squares, but this time we'll use our `tri()` function. Enter the code in Listing 5-15 in place of the `def draw()` section in Processing and then run it.

triangles.pyde

```
def setup():
    size(600,600)
    rectMode(CENTER)

t = 0

def draw():
    global t
    background(255)#white
    translate(width/2,height/2)
 ❶ for i in range(90):
        #space the triangles evenly
        #around the circle
        rotate(radians(360/90))
      ❷ pushMatrix() #save this orientation
        #go to circumference of circle
        translate(200,0)
        #spin each triangle
        rotate(radians(t))
        #draw the triangle
        tri(100)
        #return to saved orientation
      ❸ popMatrix()
    t += 0.5

def tri(length):

 ❹ noFill() #makes the triangle transparent

    triangle(0,-length,
             -length*sqrt(3)/2, length/2,
             length*sqrt(3)/2, length/2)
```

Listing 5-15: Creating 90 rotating triangles

At ❶, we use the `for` loop to arrange 90 triangles around the circle, making sure they're evenly spaced by dividing 360 by 90. Then at ❷ we use `pushMatrix()` to save this position before moving the grid around. At the end of the loop at ❸ we use `popMatrix()` to return to the saved position. In the `tri()` function at ❹, we add the `noFill()` line to make the triangles transparent.

Now we have 90 rotating transparent triangles, but they're all rotating in exactly the same way. It's kind of cool, but not as cool as Antonsen's sketch yet. Next, you'll learn how to make each triangle rotate a little differently from the adjacent ones to make the pattern more interesting.

PHASE-SHIFTING THE ROTATION

We can change the pattern in which the triangles rotate with a *phase shift*, which makes each triangle lag a little bit behind its neighbor, giving the sketch a "wave" or "cascade" effect. Each triangle has been assigned a number in the loop, represented by `i`. We need to add `i` to `t` in the `rotate(radians(t))` function, like this:

```
    rotate(radians(t+i))
```

When you run this, you should see something like Figure 5-23.

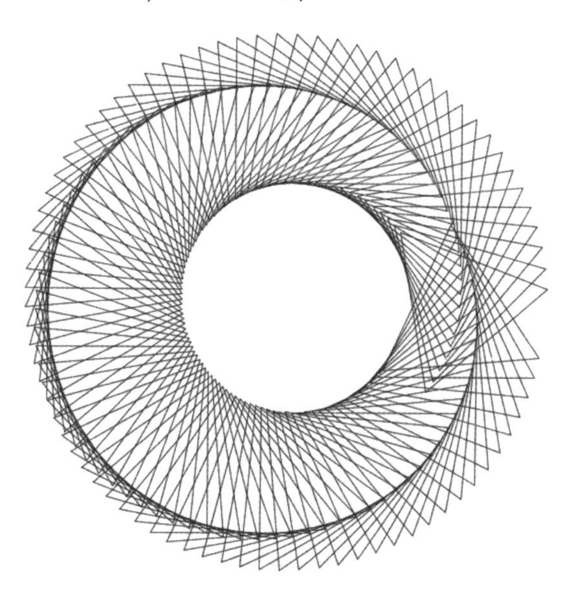

Figure 5-23: Rotating triangles with phase shift

Notice there's a break in the pattern on the right side of the screen. This break in the pattern is caused by the phase shifts not matching up from the beginning triangle to the last triangle. We want a nice, seamless pattern, so we have to make the phase shifts add up to a multiple of 360 degrees to complete the circle. Because there are 90 triangles in the design, we'll divide 360 by 90 and multiply that by `i`:

```
    rotate(radians(t+i*360/90))
```

It's easy enough to calculate 360/90, which is 4, and then use that number to plug into the code, but I'm leaving the expression in because we'll need it in case we want to change the number of triangles later. For now, this should create a nice seamless pattern, as shown in Figure 5-24.

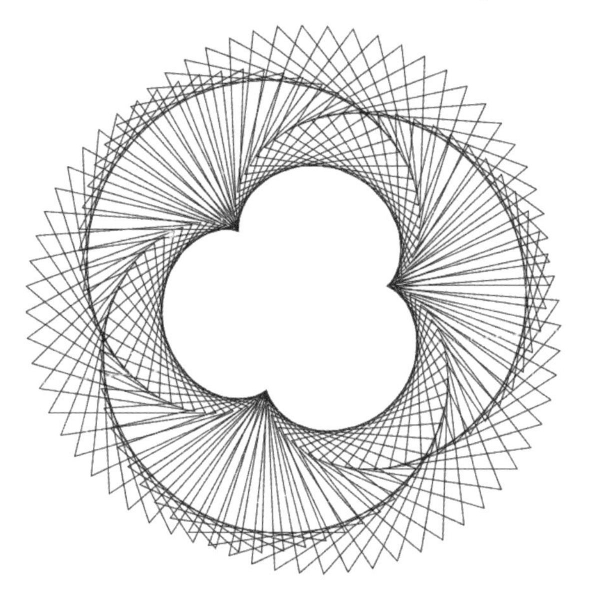

Figure 5-24: Seamlessly rotating triangles with phase shift

By making our phase shifts add up to a multiple of 360, we were able to remove the break in the pattern.

FINALIZING THE DESIGN

To make the design look more like the one in Figure 5-15, we need to change the phase shift a little. Play around with it yourself to see how you can change the look of the sketch!

Here, we're going to change the phase shift by multiplying `i` by 2, which will increase the shift between each triangle and its neighbor. Change the `rotate()` line in your code to the following:

```
        rotate(radians(t+2*i*360/90))
```

After making this change, run the code. As you can see in Figure 5-25, our design now looks very close to the design we were trying to re-create.

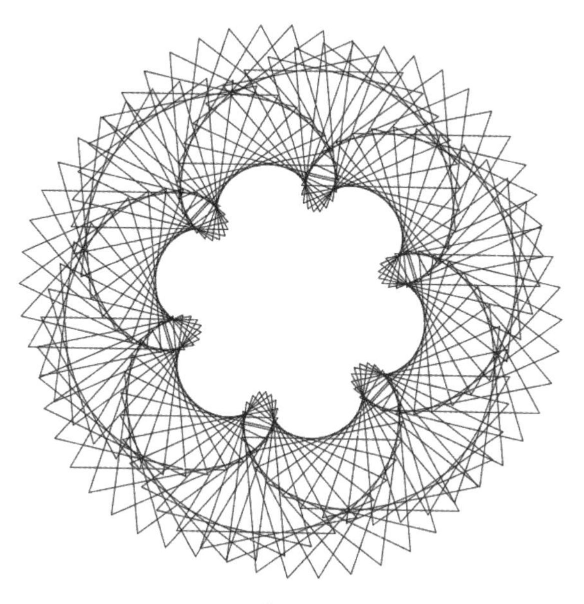

Figure 5-25: Re-creation of Antonsen's "90 Rotating Equilateral Triangles" from Figure 5-15

Now that you've learned how to re-create a complicated design like this, try the next exercise to test your transformation skills!

EXERCISE 5-2: RAINBOW TRIANGLES

Color each triangle of the rotating triangle sketch using `stroke()`. It should look like this.

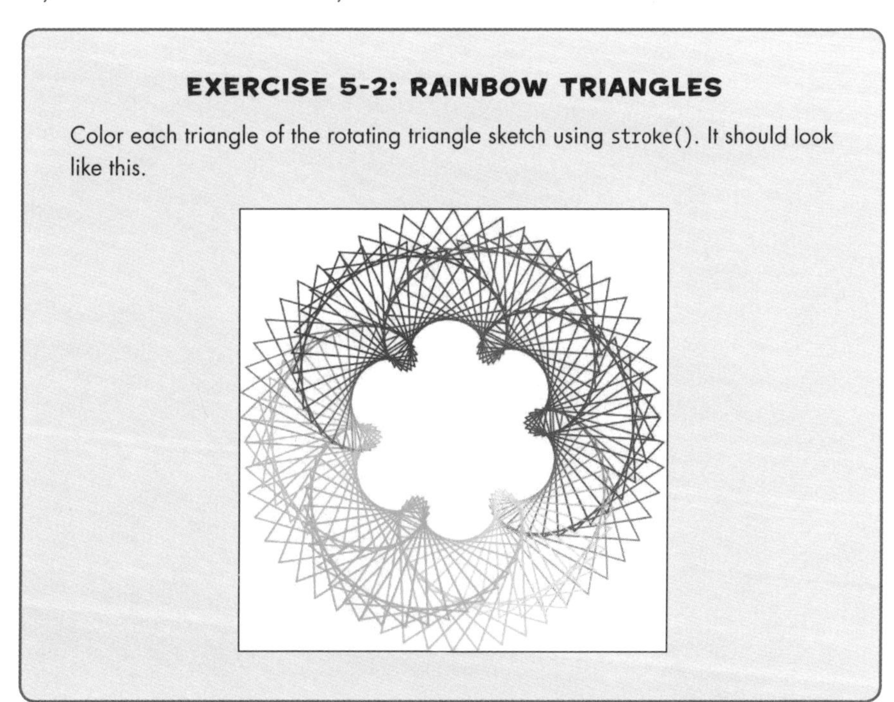

SUMMARY

In this chapter, you learned how to draw shapes like circles, squares, and triangles and arrange them into different patterns using Processing's built-in transformation functions. You also learned how to make your shapes dynamic by animating your graphics and adding color. Just like how Nasrudin's house was just a collection of bricks, the complicated code examples in this chapter are just a collection of simpler shapes or functions.

In the next chapter, you'll build on what you learned in this chapter and expand your skills to using trigonometric functions like sine and cosine. You'll draw even cooler designs and write new functions to create even more complicated behaviors, like leaving a trail and creating any shape from a bunch of vertices.

CREATING OSCILLATIONS WITH TRIGONOMETRY

I've got an oscillating fan at my house. The fan goes back and forth. It looks like the fan is saying "No." So I like to ask it questions that a fan would say "No" to. "Do you keep my hair in place? Do you keep my documents in order? Do you have three settings? Liar!" My fan lied to me.
—Mitch Hedberg

Trigonometry literally means the study of triangles. Specifically, it is the study of right triangles and the special ratios that exist between their sides. Judging from what's taught in a traditional trigonometry class, though, you'd think that's where it ends. Figure 6-1 shows just one part of a typical trigonometry homework assignment.

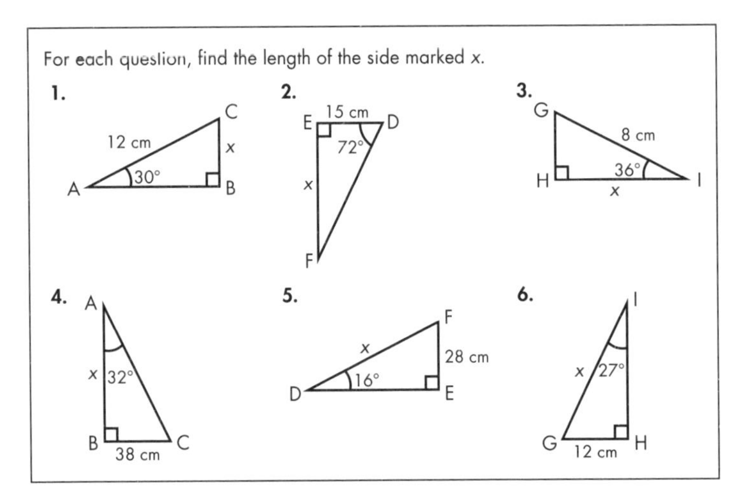

Figure 6-1: Question after question in traditional trig class on unknown sides in triangles

This is the kind of task most people remember from their trigonometry class, where solving for unknown sides in a triangle is a common assignment. *But this is seldom how trig functions are used in reality.* The more common uses of trig functions such as sine and cosine are for oscillating motion, like water, light, and sound waves. Suppose you take your graphing code from *grid.pyde* in Chapter 4 and change the function to the following:

```
def f(x):
    return sin(x)
```

In this case, you'd get this output shown in Figure 6-2.

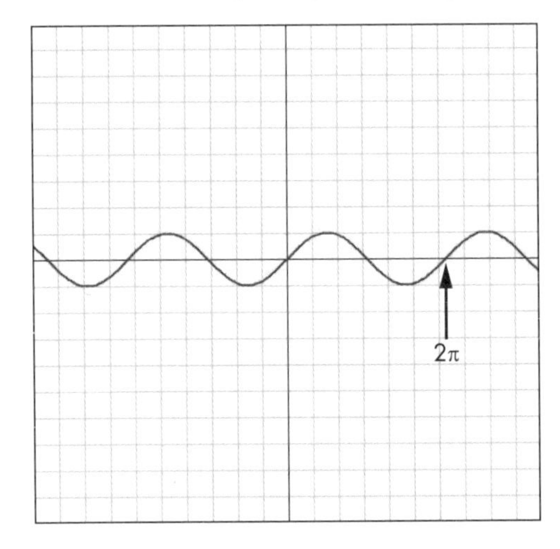

Figure 6-2: A sine wave

The values on the x-axis are the radians, the input of the sine function. The y-axis is the output. If you put `sin(1)` into your calculator or the Python shell, you'll get out a long decimal starting with 0.84. . . . That's the height of the curve when x = 1. It's almost at the top of the curve in Figure 6-2. Put `sin(3)` into the calculator and you'll get 0.14. . . . On the curve, you can see it's almost on the x-axis when x = 3. Enter any other values for x, and the output should follow this up-and-down pattern, *oscillating* between 1 and –1. The wave takes just over six units to make a complete wave, or one *wavelength*, which we also call the *period* of the function. The period of the sine function is 2π, or 6.28 radians in Processing and Python. In school, you won't go any further than drawing lots of waves like this. But in this chapter, you'll use sine, cosine, and tangent to simulate oscillating motion in real time. You'll also use trigonometry to make some interesting, dynamic, interactive sketches in Processing. The main trig functions are shown in Figure 6-3.

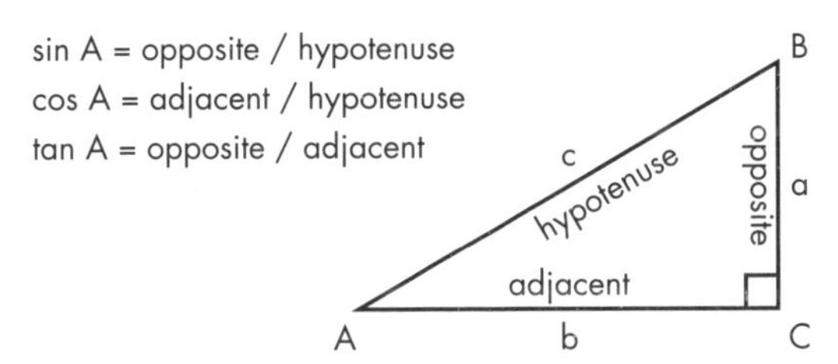

Figure 6-3: The ratios of the sides of a right triangle

We'll use trig functions to generate polygons of any number of sides as well as stars with any (odd) number of prongs. After that, you'll create a sine wave from a point rotating around a circle. You'll draw Spirograph- and harmonograph-type designs, which require trig functions. You'll also oscillate a wave of colorful points in and out of a circle!

Let's start by discussing how using trig functions is going to make transforming, rotating, and oscillating shapes much easier than before.

USING TRIGONOMETRY FOR ROTATIONS AND OSCILLATIONS

First of all, sines and cosines make rotations a cinch. In Figure 6-3, sin A is expressed as the opposite side divided by the hypotenuse, or side a divided by side c:

$$\sin A = \frac{a}{c}$$

Solve this for side a, and you get the hypotenuse times the sine of A:

$$a = c \text{ Sin } A$$

Therefore, the y-coordinate of a point can be expressed as the distance from the origin times the sine of the angle the point makes with the horizontal. Imagine a circle with radius r, the hypotenuse of the triangle, rotating around the point at (0,0), as shown in Figure 6-4.

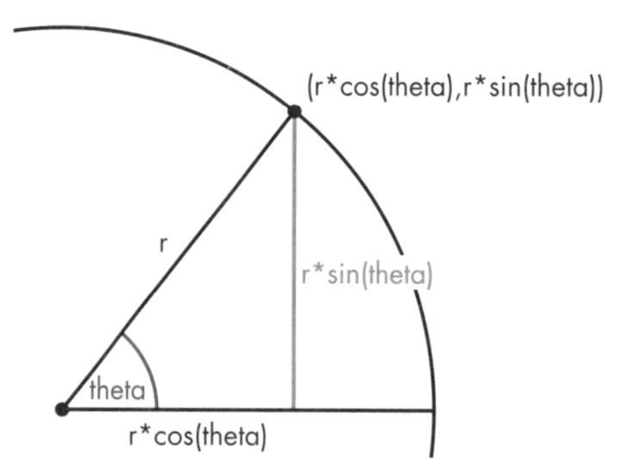

Figure 6-4: Polar form of coordinates of a point

To rotate a point, we're going to keep the radius of the circle constant and simply vary theta, the angle. The computer is going to do the hard part of recalculating all the positions of the point by multiplying the radius r by the cosine or sine of the angle theta! We also need to remember that sine and cosine expect radian input, not degrees. Fortunately, you've already learned how easy it is to use Processing's built-in `radians()` and `degrees()` functions to convert to whatever units we want.

WRITING FUNCTIONS TO DRAW POLYGONS

Thinking about vertices as points that rotate around a center makes creating polygons very easy. Recall that a polygon is a many-sided figure; a *regular polygon* is made by connecting a certain number of points equally spaced around a circle. Remember how much geometry we needed to know to draw an equilateral triangle in Chapter 5? With trigonometry functions helping us with rotations, all we have to do to draw polygons is use Figure 6-4 to create a polygon function.

Open a new sketch in Processing and save it as *polygon.pyde*. Then enter the code in Listing 6-1 to make a polygon using by the `vertex()` function.

polygon.pyde

```
def setup():
    size(600,600)

def draw():
    beginShape()
    vertex(100,100)
    vertex(100,200)
    vertex(200,200)
    vertex(200,100)
    vertex(150,50)
    endShape(CLOSE)
```

Listing 6-1: Drawing a polygon using `vertex()`

We could always draw polygons using `line()`, but once we connect all the lines, we couldn't fill in the shape with color. The Processing functions `beginShape()` and `endShape()` define any shape we want by using the `vertex()` function to say where the points of the shape should be. This lets us create as many vertices as we want.

We always start the shape with `beginShape()`, list all the points on the shape by sending them to the `vertex()` function, and finally end the shape with `endShape()`. If we put `CLOSE` inside the `endShape()` function, the program will connect the last vertex with the first vertex.

When you run this code, you should see something like Figure 6-5.

Figure 6-5: A house-shaped polygon made from vertices

However, it's laborious to enter more than four or five points manually. It would be great if we could just rotate a single point around another point using a loop. Let's try that next.

DRAWING A HEXAGON WITH LOOPS

Let's use a `for` loop to create six vertices of a hexagon using the code in Listing 6-2.

polygon.pyde

```
def draw():
    translate(width/2,height/2)
    beginShape()
    for i in range(6):
        vertex(100,100)
        rotate(radians(60))
    endShape(CLOSE)
```

Listing 6-2: Trying to use `rotate()` inside a `for` loop

However, you'll find out that if you run this code, you get a blank screen! You can't use the `rotate()` function inside a shape because this function spins the entire coordinate system. This is *precisely* why we need the sine and cosine notation you saw in Figure 6-4 to rotate the vertices!

Figure 6-6 shows how the expression `(r*cos(60*i),r*sin(60*i))` creates each vertex of a hexagon. When i = 0, the angle in the parentheses will be 0 degrees; when i = 1, the angle will be 60 degrees; and so on.

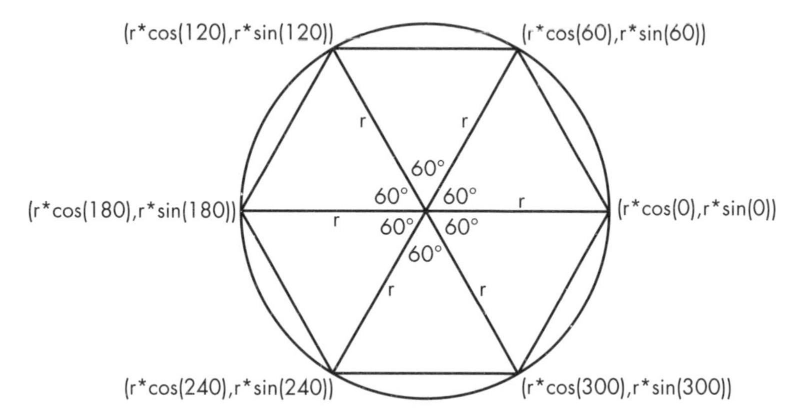

Figure 6-6: Using sines and cosines to rotate a point around the center

To re-create this hexagon in code, we have to create a variable, `r`, that represents the distance from the center of rotation to each vertex, which won't change. The only thing we need to change is the number of degrees in the `sin()` and `cos()` functions, which are all multiples of 60. Generally, it can be written like this:

```
for i in range(6):
    vertex(r*cos(60*i),r*sin(60*i))
```

First, we make `i` go from 0 to 5 so that every vertex will be a multiple of 60 (0, 60, 120, and so on), as shown in Figure 6-7. Let's change `r` to 100 and convert the degree numbers to radians so the code looks like Listing 6-3.

polygon.pyde

```
def setup():
    size(600,600)

def draw():
    translate(width/2,height/2)
    beginShape()
    for i in range(6):
        vertex(100*cos(radians(60*i)),
               100*sin(radians(60*i)))
    endShape(CLOSE)
```

Listing 6-3: Drawing a hexagon

Now that we've set `r` equal to 100 and converted the degrees to radians, when we run this code, we should see a hexagon like in Figure 6-7.

In fact, we could create a function to make *any* polygon this way!

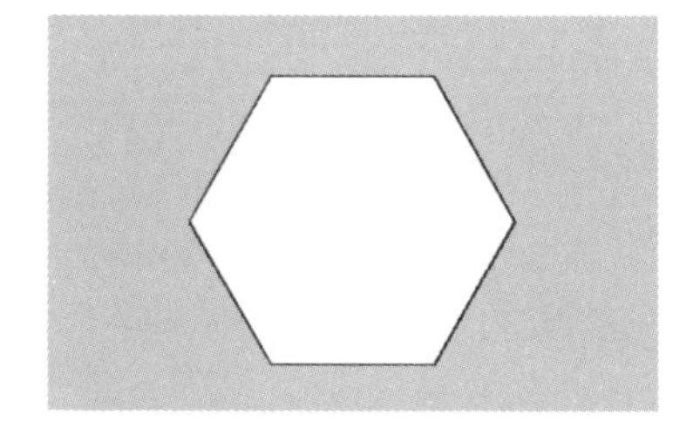

Figure 6-7: A hexagon built with a `vertex()` function and a `for` loop

DRAWING AN EQUILATERAL TRIANGLE

Now let's make an equilateral triangle using this function. Listing 6-4 shows a simpler way to make an equilateral triangle using looping instead of using square roots like we did in Chapter 5.

polygon.pyde

```
def setup():
    size(600,600)

def draw():
    translate(width/2,height/2)
    polygon(3,100) #3 sides, vertices 100 units from the center

def polygon(sides,sz):
    '''draws a polygon given the number
    of sides and length from the center'''
    beginShape()
    for i in range(sides):
    step = radians(360/sides)
        vertex(sz*cos(i * step),
               sz*sin(i * step))
    endShape(CLOSE)
```

Listing 6-4: Drawing an equilateral triangle

In this example, we create a `polygon()` function that draws a polygon given the number of sides (`sides`) and the size of the polygon (`sz`). The rotation for each vertex is 360 divided by `sides`. For our hexagon, we rotate by 60 degrees because there are six sides to a hexagon (360 / 6 = 60). The line `polygon(3,100)` calls the polygon function and passes two inputs: 3 for the number of sides and 100 for the distance from the center to the vertices.

Run this code and you should get what's shown in Figure 6-8.

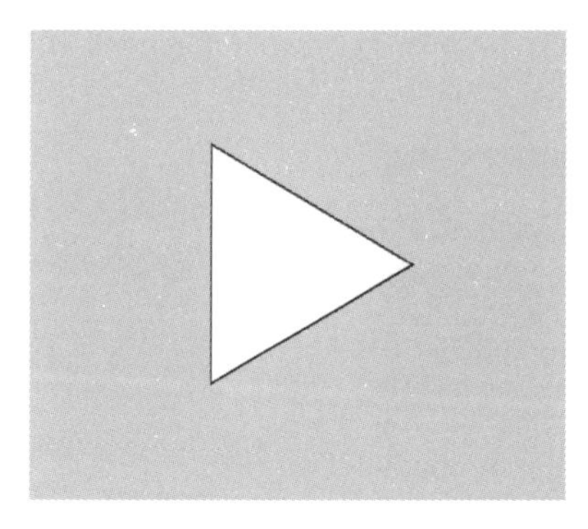

Figure 6-8: An equilateral triangle!

Now making regular polygons of any number of sides should be a breeze. No square roots necessary! Figure 6-9 shows some sample polygons you can make using the `polygon()` function.

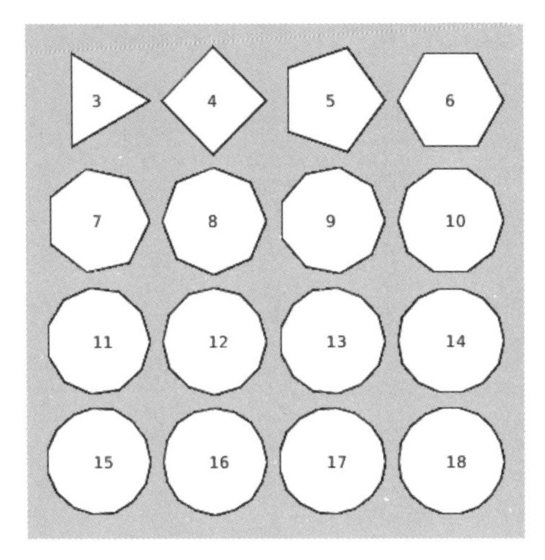

Figure 6-9: All the polygons you want!

Try updating the numbers in `polygon(3,100)` to see how the polygons change shape!

MAKING SINE WAVES

Like Mitch Hedberg's fan at the beginning of the chapter, sines and cosines are for rotating and oscillating. Sine and cosine functions make waves when the height of a point on a circle is measured over time. To make this more concrete, let's create a circle to visualize making sine waves by putting a point (shown as a red ellipse) on the circumference of the circle. As this point travels around the circle, its height over time will draw out a sine wave.

Start a new Processing sketch and save it as *CircleSineWave.pyde*. Create a big circle on the left side of the screen, like in Figure 6-10. Try it yourself before looking at the code.

Figure 6-10: The start of the sine wave sketch

Listing 6-5 shows the code to make the sketch of a red point on the circumference of a big circle.

CircleSine Wave.pyde

```
r1 = 100 #radius of big circle
r2 = 10  #radius of small circle
t = 0 #time variable

def setup():
    size(600,600)

def draw():
    background(200)
    #move to left-center of screen
    translate(width/4,height/2)
    noFill() #don't color in the circle
    stroke(0) #black outline
    ellipse(0,0,2*r1,2*r1)

    #circling ellipse:
    fill(255,0,0) #red
    y = r1*sin(t)
    x = r1*cos(t)
    ellipse(x,y,r2,r2)
```

Listing 6-5: Our circle and the point

First, we declare variables for the radii of the circles, and we use `t` to represent the time it takes to make the point move. In `draw()`, we set the background to `gray(200)`, translated to the center of the screen, and draw the big circle with radius `r1`. Next, we draw the circling ellipse by using our polar coordinates for x and y.

To make the ellipse rotate around the circle, all we have to do is vary the number inside the trig functions (in this case, `t`). At the end of the `draw()` function, we simply make the time variable go up by a little bit, like this:

```
    t += 0.05
```

If you try to run this code right now, you'll get an error message about `local variable 't' referenced before assignment`. Python functions have local variables, but we want the `draw()` function to use the global time variable `t`. Therefore, we have to add the following line to the beginning of the `draw()` function:

```
global t
```

Now you'll see a red ellipse traveling along the circumference of the circle, as in Figure 6-11.

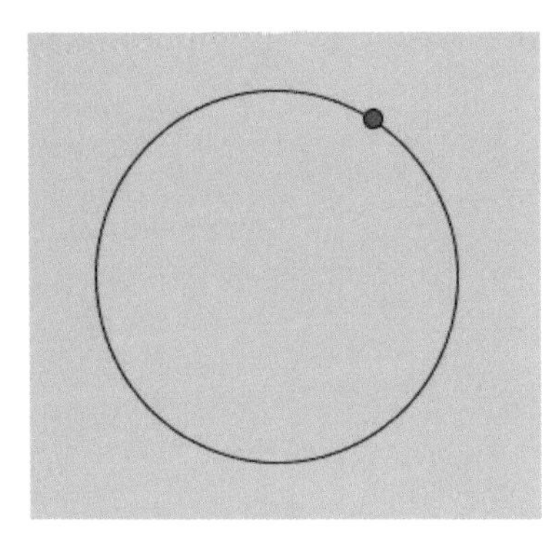

Figure 6-11: The red ellipse travels along the circumference of the big circle.

Now we need to choose a place over to the right of the screen to start drawing the wave. We'll extend a green line from the red ellipse to, say, x = 200. Add these lines to your `draw()` function right before `t += 0.05`. The full code for drawing the sine wave should look like Listing 6-6.

CircleSineWave.pyde

```
r1 = 100 #radius of big circle
r2 = 10  #radius of small circle
t = 0 #time variable

def setup():
    size(600,600)

def draw():
    global t
    background(200)
    #move to left-center of screen
    translate(width/4,height/2)
    noFill() #don't color in the circle
    stroke(0) #black outline
    ellipse(0,0,2*r1,2*r1)

    #circling ellipse:
    fill(255,0,0) #red
    y = r1*sin(t)
    x = r1*cos(t)
    ellipse(x,y,r2,r2)

    stroke(0,255,0) #green for the line
    line(x,y,200,y)
    fill(0,255,0) #green for the ellipse
    ellipse(200,y,10,10)

    t += 0.05
```

Listing 6-6: Adding a line to draw the wave

Here, we draw a green line on the same height (y-value) as the rotating red ellipse. This green line stays parallel to the horizontal, so as the red ellipse goes up and down, the green ellipse will be at the same height. When you run your program, you'll see something like Figure 6-12.

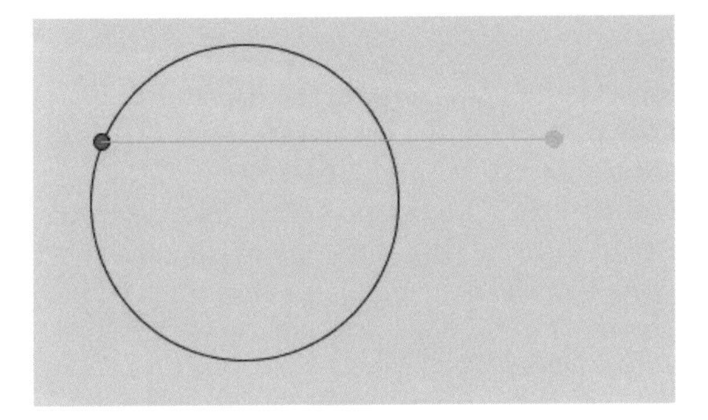

Figure 6-12: Getting ready to draw the wave!

You can see that we've added a green ellipse that only measures how far up and down the red ellipse moves, nothing else.

LEAVING A TRAIL

Now we want the green ellipse to leave a trail to show its height over time. Leaving a trail really means that we save all the heights and display them—every loop. To save a bunch of things, like numbers, letters, words, points, and so on, we need a *list*. Add this line to the variables we declared at the beginning of the program, before the `setup()` function:

```
circleList = []
```

This creates an empty list in which we'll save the locations of the green ellipse. Add the `circleList` variable to the `global` line in the `draw()` function:

```
global t, circleList
```

After we calculate x and y in the `draw()` function, we need to add the y-coordinate to the `circleList`, but there are a couple of different ways to do this. You already know the `append()` function, but this adds the point at the end of the list. We could use Python's `insert()` function to put the new points at the beginning of the list, like so:

```
circleList.insert(0,y)
```

However, the list is going to get bigger every loop. We could limit its length to 250 by adding the new value to the first 249 items already in the list, as shown in Listing 6-7.

```
    y = r1*sin(t)
    x = r1*cos(t)
```

```
    #add point to list:
    circleList = [y] + circleList[:249]
```

Listing 6-7: Adding a point to a list and limiting the list to 250 points

The new line of code concatenates the list containing the y-value we just calculated and the first 249 items in the `circleList`. That 250-point list now becomes the new `circleList`.

At the end of the `draw()` function (before incrementing `t`), we'll put in a loop that iterates over all the elements of the `circleList` and draws a new ellipse, to look like the green ellipse is leaving a trail. This is shown in Listing 6-8.

```
    #loop over circleList to leave a trail:
    for i in range(len(circleList)):
        #small circle for trail:
        ellipse(200+i,circleList[i],5,5)
```

Listing 6-8: Looping over the circle list and drawing an ellipse at each point in the list

This code uses a loop, with `i` going up from 0 to the length of the `circleList` and drawing an ellipse for each point in the list. The x-value starts at 200 and is incremented by whatever value `i` is. The y-value of the ellipse is the y-value we saved to the `circleList`.

When you run this, you'll see something like Figure 6-13.

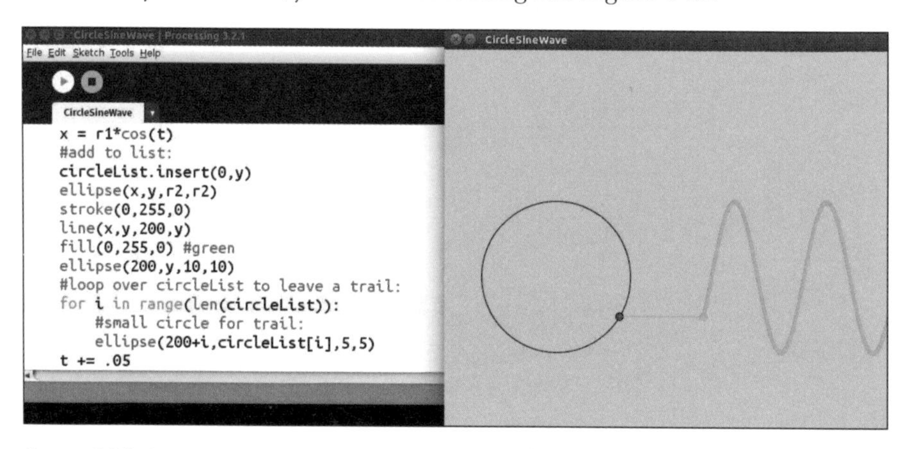

Figure 6-13: A sine wave!

You can see the wave being drawn out, leaving a green trail.

USING PYTHON'S BUILT-IN ENUMERATE() FUNCTION

You can also draw an ellipse at each point in the list using Python's built-in `enumerate()` function. It's a handy and more "Pythonic" way of keeping track of the index and value of the items in a list. To see this in action, open a new file in IDLE and enter the code in Listing 6-9.

```
>>> myList = ["I","love","using","Python"]
>>> for index, value in enumerate(myList):
        print(index,value)

0 I
1 love
2 using
3 Python
```

Listing 6-9: Learning to use Python's `enumerate()` *function*

You'll notice there are two variables (index and value) instead of just one (`i`). To use the `enumerate()` function in your circle list, you can use two variables to keep track of the iterator (`i`, the index) and the circle (`c`, the value), like in Listing 6-10.

```
    #loop over circleList to leave a trail:
    for i,c in enumerate(circleList):
        #small circle for trail:
        ellipse(200+i,c,5,5)
```

Listing 6-10: Using `enumerate()` *to get the index and the value of every item in a list*

The final code should look like what you see in Listing 6-11.

CircleSineWave.pyde

```
r1 = 100 #radius of big circle
r2 = 10  #radius of small circle
t = 0 #time variable
circleList = []

def setup():
    size(600,600)

def draw():
    global t, circleList
    background(200)
    #move to left-center of screen
    translate(width/4,height/2)
    noFill() #don't color in the circle
    stroke(0) #black outline
    ellipse(0,0,2*r1,2*r1)

    #circling ellipse:
    fill(255,0,0) #red
    y = r1*sin(t)
    x = r1*cos(t)
    #add point to list:
    circleList = [y] + circleList[:245]

    ellipse(x,y,r2,r2)
    stroke(0,255,0) #green for the line
    line(x,y,200,y)
    fill(0,255,0) #green for the ellipse
    ellipse(200,y,10,10)
```

```
    #loop over circleList to leave a trail:
    for i,c in enumerate(circleList):
        #small circle for trail:
        ellipse(200+i,c,5,5)

    t += 0.05
```

Listing 6-11: The final code for the CircleSineWave.pyde *sketch*

This is the animation that's usually shown to beginning trig students, and you've made your own version!

CREATING A SPIROGRAPH PROGRAM

Now that you know how to rotate circles and leave trails, let's make a Spirograph-type model! *Spirograph* is a toy that's made up of two overlapping circular gears that slide against each other. The gears have holes you can put pens and pencils through to draw cool, curvy designs. Many people played with Spirograph as kids, drawing the designs by hand. But we can make Spirograph-type designs using a computer and the sine and cosine code you just learned.

First, start a new sketch in Processing called *spirograph.pyde*. Then add the code in Listing 6-12.

spirograph.pyde

```
r1 = 300.0 #radius of big circle
r2 = 175.0 #radius of circle 2
r3 = 5.0   #radius of drawing "dot"
#location of big circle:
x1 = 0
y1 = 0
t = 0 #time variable
points = [] #empty list to put points in

def setup():
    size(600,600)

def draw():
    global r1,r2,x1,y1,t
    translate(width/2,height/2)
    background(255)
    noFill()
    #big circle
    stroke(0)
    ellipse(x1,y1,2*r1,2*r1)
```

Listing 6-12: Getting our big circle on the screen

We first put a big circle in the middle of the screen and create variables for the big circle, and then we put a smaller circle on its circumference, like the discs in a Spirograph set.

DRAWING THE SMALLER CIRCLE

Let's place the smaller circle on the circumference of the big circle, as in Figure 6-14.

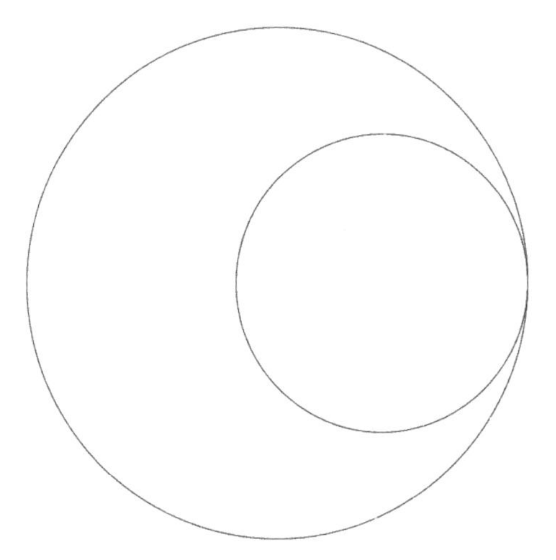

Figure 6-14: The two circles

Next, we'll make the smaller circle rotate around "inside" the bigger circle, just like a Spirograph gear. Update the code in Listing 6-12 with the code in Listing 6-13 to draw the second circle.

```
#big circle
stroke(0)
ellipse(x1,y1,2*r1,2*r1)

#circle 2
x2 = (r1 - r2)
y2 = 0
ellipse(x2,y2,2*r2,2*r2)
```

Listing 6-13: Adding the smaller circle

To make the smaller circle rotate around inside the bigger circle, we need to add the sine and cosine parts to the location of "circle 2" so it'll oscillate.

ROTATING THE SMALLER CIRCLE

Finally, at the very end of the `draw()` function, we have to increment our time variable, `t`, as in Listing 6-14.

```
#big circle
stroke(0)
ellipse(x1,y1,2*r1,2*r1)
```

```
    #circle 2
    x2 = (r1 - r2)*cos(t)
    y2 = (r1 - r2)*sin(t)
    ellipse(x2,y2,2*r2,2*r2)
    t += 0.05
```

Listing 6-14: The code to make the circle rotate

This means circle 2 will oscillate up and down, and left and right, in a circular path inside the big circle. Run the code, and you should see circle 2 spinning nicely! But how about that hole on the gear where the pen sits and draws the trail? We'll create a third ellipse to represent that point. Its location will be the second circle's center plus the difference of the radii. The code for the "drawing dot" is shown in Listing 6-15.

```
    #drawing dot
    x3 = x2+(r2 - r3)*cos(t)
    y3 = y2+(r2 - r3)*sin(t)
    fill(255,0,0)
    ellipse(x3,y3,2*r3,2*r3)
```

Listing 6-15: Adding the drawing dot

When you run this code, you'll see the drawing dot right on the edge of circle 2, rotating as if circle 2 were sliding along circle 1's circumference. Circle 3 (the drawing dot) has to be a certain proportion between the center of circle 2 and its circumference, so we need to introduce a proportion variable (`prop`) before the `setup()` function. Be sure to declare it as a global variable at the beginning of the `draw()` function, as you see in Listing 6-16.

```
prop = 0.9
--snip--

global r1,r2,x1,y1,t,prop

--snip--
x3 = x2+prop*(r2 - r3)*cos(t)
y3 = y2+prop*(r2 - r3)*sin(t)
```

Listing 6-16: Adding the proportion variable

Now we have to figure out how fast the drawing dot rotates. It only takes a little algebra to prove its angular velocity (how fast it spins around) is the ratio of the size of the big circle to the little circle. Note that the negative sign means the dot spins in the opposite direction. Change the `x3` and `y3` lines in the `draw()` function to this:

```
x3 = x2+prop*(r2 - r3)*cos(-((r1-r2)/r2)*t)
y3 = y2+prop*(r2 - r3)*sin(-((r1-r2)/r2)*t)
```

All that's left is to save the dot (x3,y3) to a points list and draw lines between the points, just like we did in the wave sketch. Add the points list to the global line:

```
global r1,r2,x1,y1,t,prop,points
```

After drawing the third ellipse, put the points into a list. This is the same procedure we used in *CircleSineWave.pyde* earlier in the chapter. Finally, go through the list and draw lines between the points, as in Listing 6-17.

```
    fill(255,0,0)
    ellipse(x3,y3,2*r3,2*r3)
    #add points to list
    points = [[x3, y3]] + points[:2000]
    for i,p in enumerate(points): #go through the points list
        if i < len(points)-1: #up to the next to last point
            stroke(255,0,0) #draw red lines between the points
            line(p[0],p[1],points[i+1][0],points[i+1][1])

    t += 0.05
```

Listing 6-17: Graphing the points in the Spirograph

We used a similar trick for adding the points to the list in the circular wave example. We concatenated a list with the current point in it to a list with 2000 of the items in the circleList. This automatically limits the number of points we're saving to the points list. Run this code and watch the program draw a Spirograph, as shown in Figure 6-15.

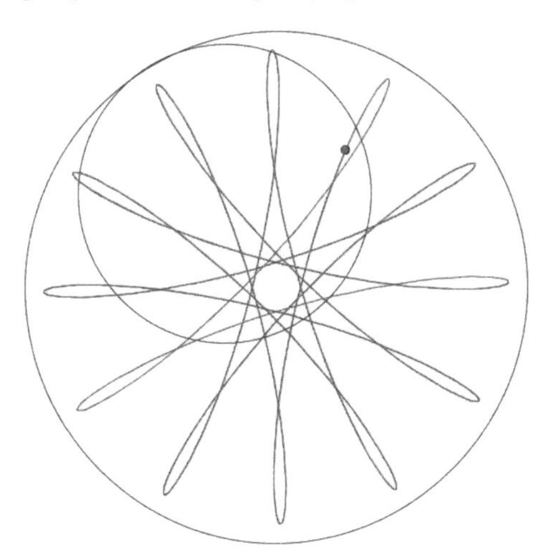

Figure 6-15: Drawing the Spirograph

You can change the size of the second circle (r2) and the position of the drawing dot (prop) to draw different designs. For example, the Spirograph in Figure 6-16 has r2 equal to 105 and prop equal to 0.8.

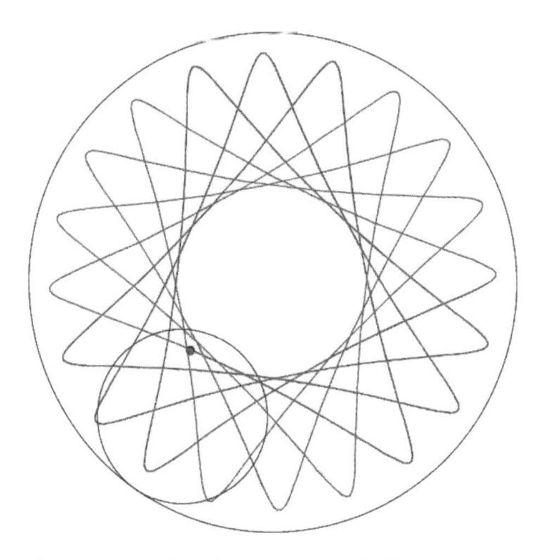

Figure 6-16: Another Spirograph design, created by changing `r2` *and* `prop`

So far, we've been making shapes oscillate up and down, or left and right, using sine and cosine, but what about making shapes oscillate in two different directions? We'll try that next.

MAKING HARMONOGRAPHS

In the 1800s, there was an invention called the *harmonograph* that was a table connected to two pendulums. When the pendulums swung, the attached pen would draw on a piece of paper. As the pendulums swung back and forth and died down (*decayed*), the patterns would change in interesting ways, as illustrated in Figure 6-17.

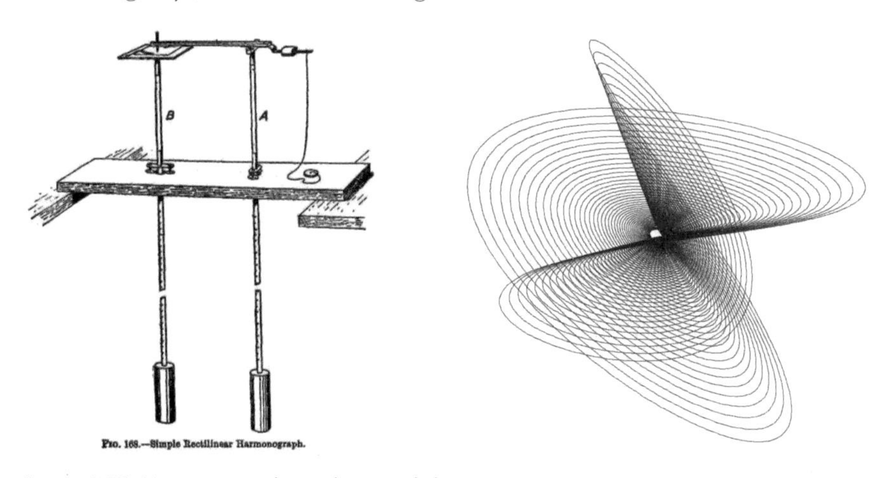

Figure 6-17: Harmonograph machine and design

Using programming and a few equations, we can model how a harmonograph draws its patterns. The equations to model the oscillation of one pendulum are

$$x = a * \cos(ft + p)e^{-dt}$$
$$y = a * \sin(ft + p)e^{-dt}$$

In these equations, *x* and *y* represent the horizontal and vertical displacement left/right and up/down distance) of the pen, respectively. Variable *a* is the amplitude (size) of the motion, *f* is the frequency of the pendulum, *t* is the elapsed time, *p* is the phase shift, e is the base of the natural logarithms (it's a constant, around 2.7), and *d* is the decay factor (how fast the pendulum slows down). The time variable, *t*, will of course be the same in both of these equations, but all the other variables can be different: the left/right frequency can be different from the up/down frequency, for example.

WRITING THE HARMONOGRAPH PROGRAM

Let's create a Python-Processing sketch that models the movement of a pendulum. Create a new Processing sketch and call it *harmonograph.pyde*. The initial code is shown in Listing 6-18.

harmonograph.pyde

```
t = 0

def setup():
    size(600,600)
    noStroke()

def draw():
    global t
 ❶ a1,a2 = 100,200 #amplitudes
    f1,f2 = 1,2 #frequencies
    p1,p2 = 0,PI/2 #phase shifts
    d1,d2 = 0.02,0.02 #decay constants
    background(255)
    translate(width/2,height/2)
 ❷ x = a1*cos(f1*t + p1)*exp(-d1*t)
    y = a2*cos(f2*t + p2)*exp(-d2*t)
    fill(0) #black
    ellipse(x,y,5,5)
    t += .1
```

Listing 6-18: The initial code for the harmonograph sketch

This is just the usual `setup()` and `draw()` functions with a time variable (`t`) and values for the amplitude (`a1,a2`), frequency (`f1,f2`), phase shift (`p1,p2`), and decay constants (`d1,d2`).

Then, starting at ❶, we define a bunch of variables to plug into the two formulas for the location of the harmonograph drawing pen. The `x =` and `y =` lines ❷ use those variables and calculate the coordinates for the ellipse.

Now run this code, and you should see the circle moving, but what is it drawing? We need to put the points in a list and then graph all the points in the list. Right after declaring the t variable, create a list called points. The code so far is shown in Listing 6-19.

harmonograph.pyde

```
t = 0
points = []

def setup():
    size(600,600)
    noStroke()

def draw():
    global t,points
    a1,a2 = 100,200
    f1,f2 = 1,2
    p1,p2 = 0,PI/2
    d1,d2 = 0.02,0.02
    background(255)
    translate(width/2,height/2)
    x = a1*cos(f1*t + p1)*exp(-d1*t)
    y = a2*cos(f2*t + p2)*exp(-d2*t)
    #save location to points List
    points.append([x,y])
    #go through points list and draw lines between them
    for i,p in enumerate(points):
        stroke(0) #black
        if i < len(points) - 1:
            line(p[0],p[1],points[i+1][0],points[i+1][1])
    t += .1
```

Listing 6-19: The code to draw a harmonograph using lines between points

We start by defining the points list at the top of the file and adding points to the global variables in the draw() function. After calculating where x and y are, we add the line to add the point [x,y] to the points list. Finally, we go through the points list and draw a line from each point to the next one. Then we use Python's enumerate() function and stop one point before the last one. This is so we don't get an error message telling us the index is out of range when it tries to draw a line from the last point to the next one. Now when we run the code, we see the dot leave a trail behind it, as in Figure 6-18.

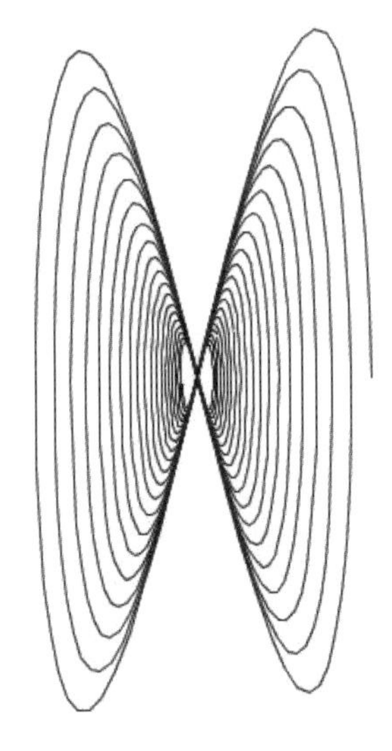

Figure 6-18: The harmonograph

Notice if you comment out the decay part of the formulas, like this, the program will simply draw over the same lines:

```
    x = a1*cos(f1*t + p1)#*exp(-d1*t)
    y = a2*cos(f2*t + p2)#*exp(-d2*t)
```

The decay models the gradual decrease in a pendulum's maximum amplitude, and it's what creates the "scalloped" effect of so many harmonograph images. The first few times it's cool to watch the code draw the design, but it takes a while. What if we could fill the `points` list all at once?

FILLING THE LIST INSTANTLY

Instead of drawing the whole list at every frame, let's come up with a way to fill the list instantly. We can cut the whole harmonograph code out of the `draw()` function and paste it into its own function, like in Listing 6-20.

```
def harmonograph(t):
    a1,a2 = 100,200
    f1,f2 = 1,2
    p1,p2 = PI/6,PI/2
    d1,d2 = 0.02,0.02
    x = a1*cos(f1*t + p1)*exp(-d1*t)
    y = a2*cos(f2*t + p2)*exp(-d2*t)
    return [x,y]
```

Listing 6-20: Separating out the `harmonograph()` function

Now in the `draw()` function, you just need a loop where you add a bunch of points for values of `t`, as in Listing 6-21.

```
def draw():
    background(255)
    translate(width/2,height/2)
    points = []
    t = 0
    while t < 1000:
        points.append(harmonograph(t))
        t += 0.01

    #go through points list and draw lines between them
    for i,p in enumerate(points):
        stroke(0) #black
        if i < len(points) - 1:
            line(p[0],p[1],points[i+1][0],points[i+1][1])
```

Listing 6-21: The new `draw()` function, which calls the `harmonograph()` function

Run this code and you'll instantly see a complete harmonograph! Because we changed the size of the ellipses and the phase shifts, this one looks different, as you can see in Figure 6-19. Change each of the values yourself and see how this changes the design!

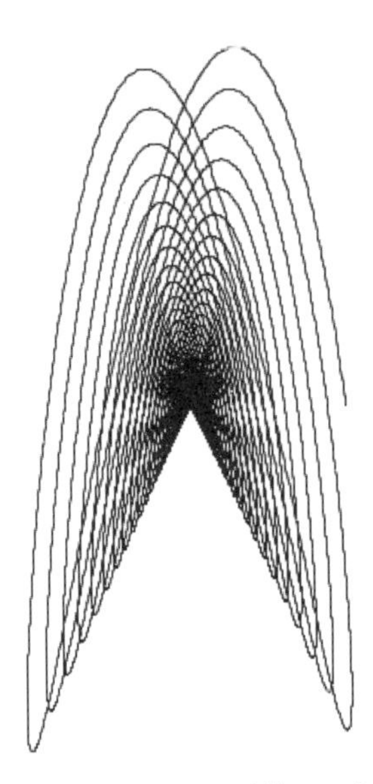

Figure 6-19: Using a different formula to make the harmonograph

TWO PENDULUMS ARE BETTER THAN ONE

We can add another pendulum to make more complicated designs by adding another term to each formula, like this:

```
x = a1*cos(f1*t + p1)*exp(-d1*t) + a3*cos(f3*t + p3)*exp(-d3*t)
y = a2*sin(f2*t + p2)*exp(-d2*t) + a4*sin(f4*t + p4)*exp(-d4*t)
```

All this does is add identical code to each line, with a few numbers changed, to simulate more than one pendulum in each direction. Of course, you have to create more variables and give them values. In Listing 6-22 are my suggestions for copying one of the designs I found at *http://www.walkingrandomly.com/?p=151.*

```
def harmonograph(t):
    a1=a2=a3=a4 = 100
    f1,f2,f3,f4 = 2.01,3,3,2
    p1,p2,p3,p4 = -PI/2,0,-PI/16,0
    d1,d2,d3,d4 = 0.00085,0.0065,0,0
    x = a1*cos(f1*t + p1)*exp(-d1*t) + a3*cos(f3*t + p3)*exp(-d3*t)
    y = a2*sin(f2*t + p2)*exp(-d2*t) + a4*sin(f4*t + p4)*exp(-d4*t)
    return [x,y]
```

Listing 6-22: The harmonograph code for the design in Figure 6-20

In Listing 6-22, all we changed were the constants for `a`, `f`, `p`, and `d` to make a completely different design. If you add `stroke(255,0,0)` to the code before drawing the lines, you'll make the lines red, as shown in Figure 6-20.

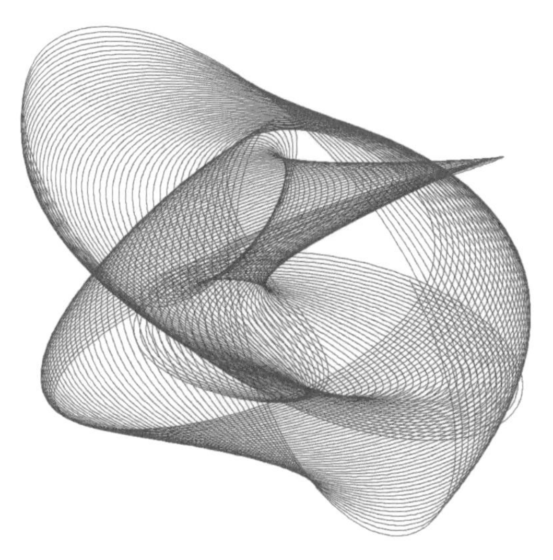

Figure 6-20: A complete harmonograph!

Listing 6-23 shows the final code for *harmonograph.pyde*.

harmonograph.pyde

```
t = 0
points = []

def setup():
    size(600,600)
    noStroke()

def draw():
    background(255)
    translate(width/2,height/2)
    points = []
    t = 0
    while t < 1000:
        points.append(harmonograph(t))
        t += 0.01

    #go through points list and draw lines between them
    for i,p in enumerate(points):
        stroke(255,0,0) #red
        if i < len(points) - 1:
            line(p[0],p[1],points[i+1][0],points[i+1][1])

def harmonograph(t):
    a1=a2=a3=a4 = 100
    f1,f2,f3,f4 = 2.01,3,3,2
    p1,p2,p3,p4 = -PI/2,0,-PI/16,0
    d1,d2,d3,d4 = 0.00085,0.0065,0,0
    x = a1*cos(f1*t + p1)*exp(-d1*t) + a3*cos(f3*t + p3)*exp(-d3*t)
    y = a2*sin(f2*t + p2)*exp(-d2*t) + a4*sin(f4*t + p4)*exp(-d4*t)
    return [x,y]
```

Listing 6-23: The final code for the harmonograph sketch

SUMMARY

Students in trigonometry class have to solve for unknown side lengths or angle measurements in triangles. But now you know the *real* use of sines and cosines is to rotate and transform points and shapes to make Spirograph and harmonograph designs! In this chapter, you saw how useful it is to save points to a list and then loop through the list to draw lines between the points. We also revisited some Python tools like `enumerate()` and `vertex()`.

In the next chapter, we'll use sines and cosines and the rotation ideas you learned in this chapter to invent a whole new kind of number! We'll also rotate and transform grids using these new numbers, and we'll create complex (pun intended) works of art using the locations of pixels!

7 COMPLEX NUMBERS

Imaginary numbers are a fine and wonderful refuge of the divine spirit, almost an amphibian between being and non-being.
—Gottfried Leibniz

Numbers containing the square root of –1 have been given a bad name in math classes. We call the square root of –1 an *imaginary number*, or *i*. Calling something "imaginary" makes it seem like it doesn't exist or like there's no real purpose for it. But imaginary numbers *do* exist, and they have a lot of real-world applications in fields such as electromagnetism, for instance.

In this chapter, you get a taste of the kinds of beautiful art you can create using *complex numbers*, which are numbers that have both a real and imaginary part written in the form of $a + bi$, where a and b are real numbers and i is the imaginary number. Because a complex number holds two different bits of information, real and imaginary, you can use it to turn one-dimensional objects into two-dimensional ones. Using Python, manipulating these numbers becomes easier, and we can use them for some very magical purposes. In fact, we use complex numbers to explain behaviors of electrons and photons, and what we think of as natural, "normal" numbers are actually complex numbers whose imaginary parts equal zero!

We begin this chapter by reviewing how to plot complex numbers in the complex coordinate plane. You also learn how to express complex numbers as Python lists and then write functions to add and multiply them. Finally, you learn how to find the magnitude, or absolute value, of a complex number. Knowing how to manipulate complex numbers will come in handy when we write the programs for creating the Mandelbrot set and the Julia set later in this chapter.

THE COMPLEX COORDINATE SYSTEM

As Frank Farris summed up in his brilliant and beautifully illustrated book *Creating Symmetry*, "Complex numbers . . . are simply a way to express the Cartesian ordered pair of real numbers, (x, y), compactly as a single number $z = x + iy$." We all know the Cartesian coordinate system uses x to represent the horizontal axis and y to represent the vertical axis, but we never add or multiply those numbers; they just represent a location.

In contrast, complex numbers not only can represent locations but they can also be operated on like any other numbers. It helps to look at complex numbers geometrically. Let's change our coordinate system a little so that now the real numbers are on the horizontal axis and the imaginary numbers are on the vertical axis, as in Figure 7-1.

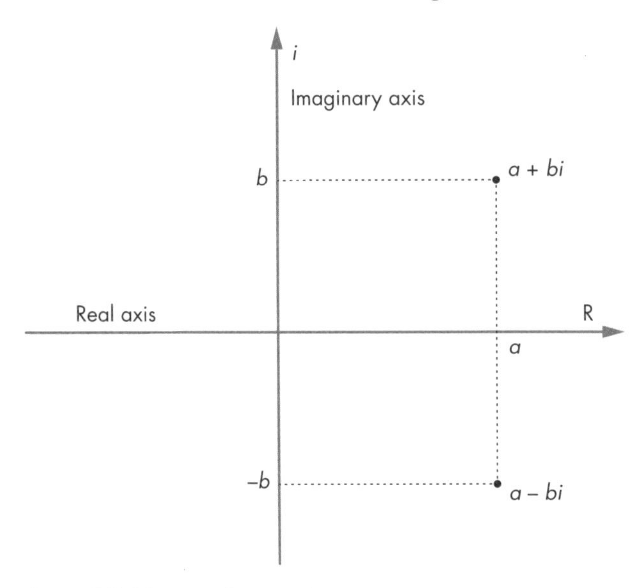

Figure 7-1: The complex coordinate system

Here, you can see where $a + bi$ and $a - bi$ would be located on a complex coordinate system.

ADDING COMPLEX NUMBERS

Adding and subtracting complex numbers are the same as with real numbers: you start at one number and take the number of steps represented by the other number. For example, to add the numbers $2 + 3i$ and $4 + i$, you would simply add the real parts and the imaginary parts of the numbers separately to get $6 + 4i$, as shown in Figure 7-2.

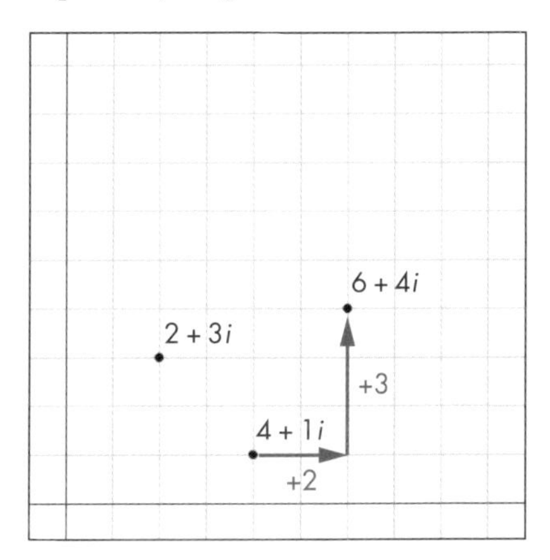

Figure 7-2: Adding two complex numbers

As you can see, we start at $4 + i$. To add $2 + 3i$, we move two units in the positive real direction and three units in the positive imaginary direction, and end up at $6 + 4i$.

Let's write the function for adding two complex numbers using the code in Listing 7-1. Open a new file in IDLE and name it *complex.py*.

```
def cAdd(a,b):
    '''adds two complex numbers'''
    return [a[0]+b[0],a[1]+b[1]]
```

Listing 7-1: The function for adding two complex numbers

Here, we define the function called `cAdd()`, giving it two complex numbers in list form `[x,y]`, which returns another list. The first term of the list, `a[0]+b[0]`, is the sum of the first terms of the complex numbers (index 0) we provide. The second term, `a[1]+b[1]`, is the sum of the second terms (index 1) of the two complex numbers. Save and run this program.

Now let's test the program using the complex numbers `u = 1 + 2i` and `v = 3 + 4i`. Plug them into our `cAdd()` function in the interactive shell, like this:

```
>>> u = [1,2]
>>> v = [3,4]
>>> cAdd(u,v)
[6, 4]
```

You should get $6 + 4i$, which is the sum of the complex numbers $1 + 2i$ and $3 + 4i$. Adding complex numbers is just like taking steps in the x-direction and then in the y-direction, and we'll see this function again when we want to create beautiful designs like the Mandelbrot set and the Julia set.

MULTIPLYING A COMPLEX NUMBER BY I

But adding complex numbers isn't the most useful thing. Multiplying them is. For example, multiplying a complex number by *i* rotates the complex number around the origin by 90 degrees. In the complex coordinate system, multiplying a real number by –1 would rotate it 180 degrees around the origin, as shown in Figure 7-3.

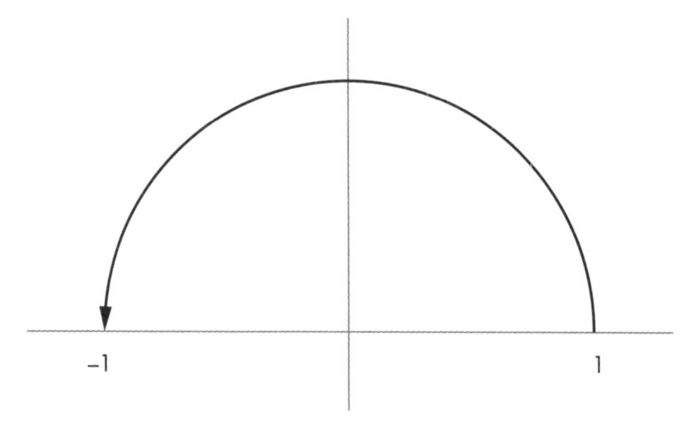

Figure 7-3: Multiplying a number by –1 as a 180-degree rotation

As you can see, 1 times –1 equals –1, which rotates 1 over to the other side of zero.

Because multiplying a complex number by –1 is the same as a 180 degree rotation, the square root of –1 would represent a 90 degree rotation, as shown in Figure 7-4.

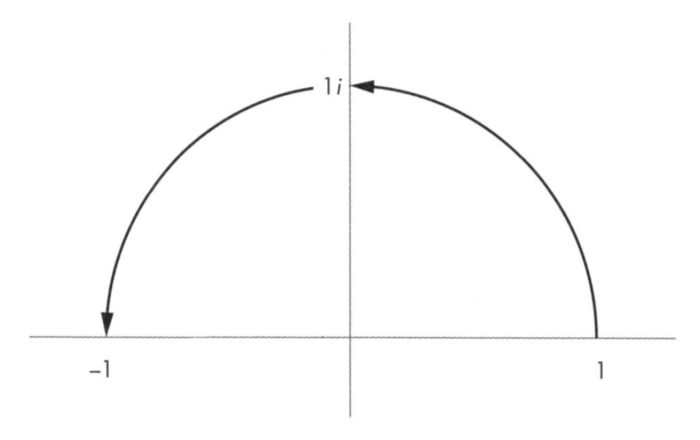

Figure 7-4: Multiplying a number by i as a 90-degree rotation

This means that i represents the square root of –1, the number that rotates us halfway to –1 when we multiply it by 1. Multiplying the result (i) by i again causes us to rotate 90 degrees more, and we end up with –1. This confirms the definition of the square root because we are able to get to the negative value of a number by multiplying it by the same number (i) twice.

MULTIPLYING TWO COMPLEX NUMBERS

Let's see what happens when we multiply two complex numbers. Just like you would multiply two binomial expressions, you can multiply two complex numbers algebraically using the FOIL method:

$$
\begin{aligned}
&(a + bi)(c + di) \\
&= ac + adi + bci + bdi^2 \\
&= ac + (ad + bc)i + bd(-1) \\
&= ac - bd + (ad + bc)i \\
&= [ac - bd, ad + bc]
\end{aligned}
$$

To make this easier, let's translate this process into a `cMult()` function, as shown in Listing 7-2.

```
def cMult(u,v):
    '''Returns the product of two complex numbers'''
    return [u[0]*v[0]-u[1]*v[1],u[1]*v[0]+u[0]*v[1]]
```

Listing 7-2: Writing the function for multiplying two complex numbers

To test the `cMult()` function, try multiplying u = $1 + 2i$ by v = $3 + 4i$. Enter the following in the interactive shell:

```
>>> u = [1,2]
>>> v = [3,4]
>>> cMult(u,v)
[-5, 10]
```

As you can see, the product is $-5 + 10i$.

Recall from the previous section that multiplying a complex number by i is the same as performing a 90 degree rotation about the origin of the complex coordinate system. Let's try it now with v = $3 + 4i$:

```
>>> cMult([3,4],[0,1])
[-4, 3]
```

The result is $-4 + 3i$. When we graph $3 + 4i$ and $-4 + 3i$, you should see something like what's shown in Figure 7-5.

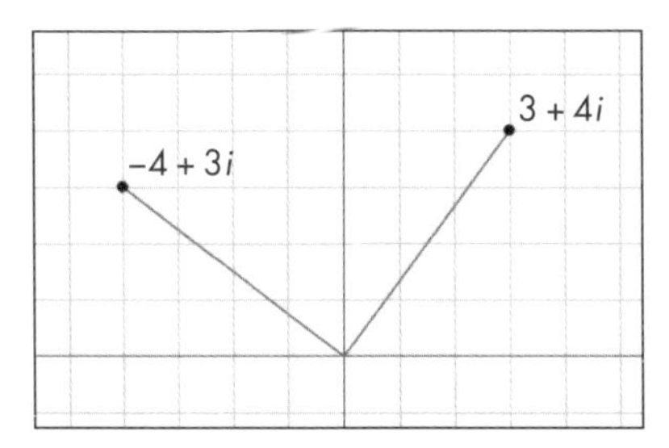

Figure 7-5: Rotating a complex number 90 degrees by multiplying by i

As you can see, $-4 + 3i$ is 90 degrees rotation from $3 + 4i$.

Now that you know how to add and multiply complex numbers, let's go over how to find the magnitude of a complex number, which you'll use to create the Mandelbrot set and Julia set.

WRITING THE MAGNITUDE() FUNCTION

The *magnitude*, or *absolute value*, of a complex number is how far the complex number is away from the origin on the complex coordinate plane. Now let's create a magnitude function using the Pythagorean theorem. Return to *complex.py* and make sure to import the square root function from Python's `math` module at the top of the file:

```
from math import sqrt
```

The `magnitude()` function is just the Pythagorean theorem:

```
def magnitude(z):
    return sqrt(z[0]**2 + z[1]**2)
```

Let's find the magnitude of the complex number 2 + i:

```
>>> magnitude([2,1])
2.23606797749979
```

Now you're ready to write a Python program that colors the pixels on the display window according to how large the complex numbers get. The unexpected behavior of complex numbers will result in an infinitely complicated design that's impossible to replicate without a computer!

CREATING THE MANDELBROT SET

To create the Mandelbrot set, we're going to represent each pixel on our display window as a complex number, z, then repeatedly square the value, and add the original number z.

$$z_{n+1} = z_n^{\ 2} + c$$

Then, we're going to do the same to the output, again and again. If the number keeps getting larger, we'll color the pixel corresponding to the original complex number according to how many iterations it takes for its magnitude to get bigger than a certain number, like 2. If the number keeps getting smaller, we'll give it a different color.

You already know that multiplying a number by a number larger than 1 makes the original number larger. A number multiplied by 1 stays the same, and multiplying by a number smaller than 1 makes the original number smaller. Complex numbers follow a similar pattern, which you can represent on the complex plane as shown in Figure 7-6.

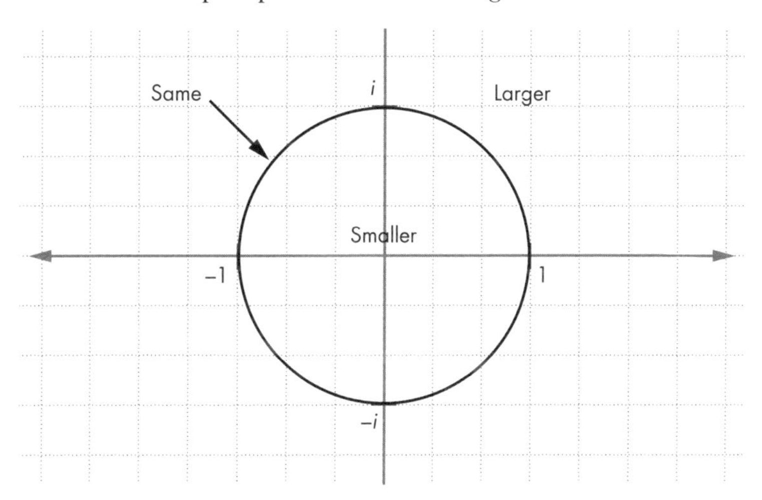

Figure 7-6: Visualizing what happens when you multiply complex numbers

If we were only multiplying complex numbers, the Mandelbrot set would look like Figure 7-6, a circle. But not only is the complex number squared, a number is added afterward. This will change the circle into an infinitely complicated and surprisingly beautiful figure. But before we can do that, we need to operate on every point on the grid!

Depending on the result of the operation, some will get smaller and *converge* to zero. Others will get bigger and *diverge*. Getting close to a number in math terms is called *converging*. Getting too big in math terms is called *diverging*. For our purposes, we'll color every pixel on the grid according to how many iterations it takes it to get too big and fly off the grid. The formula we plug the number into is similar to our `cMult()` function from Listing 7-2, with an extra step. We square the number, add the original complex number to the square, and then repeat that process until it diverges. If the magnitude of the squared complex number gets larger than 2, it means it has diverged (we can pick any number we want to be the maximum). If it never gets bigger than 2, we'll leave it black.

For example, let's try the Mandelbrot set operation manually using the complex number $z = 0.25 + 1.5i$:

```
>>> z = [0.25,1.5]
```

We square z by multiplying it by itself and saving the result to the variable z2:

```
>>> z2 = cMult(z,z)
>>> z2
[-2.1875, 0.75]
```

Then we add z2 and z using the cAdd() function:

```
>>> cAdd(z2,z)
[-1.9375, 2.25]
```

We have a function we can use to test if this complex number is more than two units away from the origin using the Pythagorean theorem. Let's use our magnitude() function from earlier to check if the magnitude of the complex number we got is greater than 2:

```
>>> magnitude([-1.9375,2.25])
2.969243380054926
```

We set the rule as follows: "If a number gets more than two units away from the origin, it diverges." Therefore, the complex number $z = 0.25 + 1.5i$ diverges after only 1 iteration!

This time, let's try with $z = 0.25 + 0.75i$, as shown next:

```
>>> z = [0.25,0.75]
>>> z2 = cMult(z,z)
>>> z3 = cAdd(z2,z)
>>> magnitude(z3)
1.1524430571616109
```

Here, we repeated the same process as before, except this time we need to add z2 and z again, saving it as z3. It's still within two units of the origin, so let's replace z with this new value and put it back through the process again. First, we create a new variable, z1, that we can use to square the original z:

```
>>> z1 = z
```

Let's repeat the process using the newest value of our complex number, z3. We'll square it, add z1, and find the magnitude:

```
>>> z2 = cMult(z3,z3)
>>> z3 = cAdd(z2,z1)
>>> magnitude(z3)
0.971392565148097
```

Because 0.97 is smaller than 1.152, we might guess that the result is getting smaller and therefore doesn't look like it's going to diverge, but we've only repeated the process twice. Doing this by hand is laborious! Let's automate the steps so that we can repeat the process quickly and easily. We'll use the squaring, adding, and finding the magnitude functions to write a function called `mandelbrot()` that automates the checking process so that we can visually separate the diverging numbers from the converging ones. What design do you think it'll make? A circle? An ellipse? Let's find out!

WRITING THE MANDELBROT() FUNCTION

Let's open a Processing sketch and call it *mandelbrot.pyde*. The Mandelbrot set we're trying to re-create here is named after the mathematician Benoit Mandelbrot, who first explored this process using computers in the 1970s. We'll repeat the squaring and adding process a maximum number of times, or until the number diverges, as shown in Listing 7-3.

```
def mandelbrot(z,num):
    '''runs the process num times
    and returns the diverge count '''
 ❶ count=0
    #define z1 as z
    z1=z
    #iterate num times
 ❷ while count <= num:
        #check for divergence
        if magnitude(z1) > 2.0:
        #return the step it diverged on
            return count
        #iterate z
      ❸ z1=cAdd(cMult(z1,z1),z)
        count+=1
    #if z hasn't diverged by the end
    return num
```

Listing 7-3: Writing the `mandelbrot()` function to check how many steps a complex number takes to diverge

The `mandelbrot()` function takes a complex number, `z`, and a number of iterations as parameters. It returns the number of times it took for `z` to diverge, and if it never diverges, it returns `num` (at the end of the function). We create a `count` variable ❶ to keep track of the iterations, and we create a new complex number, `z1`, that gets squared and so on without changing `z`.

We start a loop to repeat the process while the `count` variable is less than `num` ❷. Inside the loop we check the magnitude of `z1` to see whether `z1` has diverged, and if it has, we return `count` and stop the code. Otherwise, we square `z1` and add `z` to it ❸, which is the definition of our operation on complex numbers. Finally, we increment the `count` variable by 1 and loop through the process again.

Using the *mandelbrot.pyde* program, we can plug in our complex number $z = 0.25 + 0.75i$ and check the magnitude after every iteration. Here are the magnitudes after each loop:

```
0.7905694150420949
1.1524430571616109
```

```
0.971392565148097
1.1899160852817983
2.122862368187107
```

The first number is the magnitude of z = 0.25 + 0.75*i* before any iterations:

$$\sqrt{0.25^2 + 0.75^2} = 0.790569\ldots$$

You can see that it diverges after four iterations because it gets bigger than two units away from the origin. Figure 7-7 graphs each step so you can visualize them.

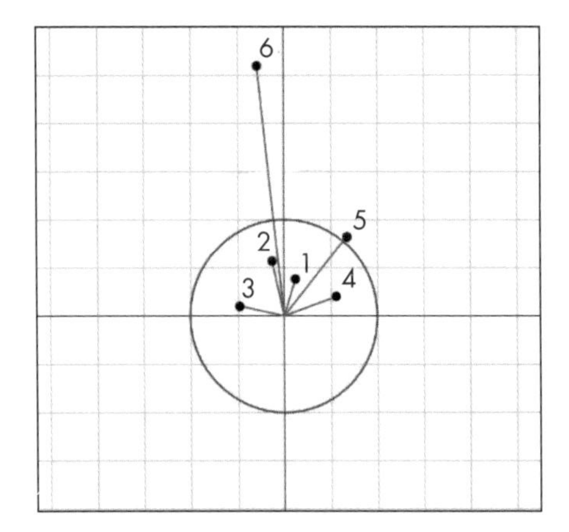

*Figure 7-7: Running the complex number 0.25 + 0.75*i *through the* `mandelbrot()` *function until it diverges*

The red circle has a radius of two units and represents the limit we put on the complex number diverging. When squaring and adding in the original value of z, we cause the locations of the numbers to rotate and translate and eventually to get further away from the origin than our rule allows.

Let's use some of the graphing tricks we learned in Chapter 4 to graph points and functions in the Processing display. Copy and paste all the complex number functions from *complex.py* (`cAdd`, `cMult`, and `magnitude`) to the bottom of *mandelbrot.pyde*. We'll use Processing's `println()` function to print to the console the number of steps it takes a point to diverge. Add the code in Listing 7-4 before the `mandelbrot()` code you wrote in Listing 7-3.

mandelbrot.pyde

```
#range of x-values
xmin = -2
xmax = 2

#range of y-values
ymin = -2
ymax = 2
```

```
#calculate the range
rangex = xmax - xmin
rangey = ymax - ymin

def setup():
    global xscl, yscl
    size(600,600)
    noStroke()
    xscl = float(rangex)/width
    yscl = float(rangey)/height

def draw():
    z = [0.25,0.75]
    println(mandelbrot(z,10))
```

Listing 7-4: The beginning of the Mandelbrot code

We calculate the range of real values (`x`) and imaginary values (`y`) at the top of the program. Inside `setup()`, we calculate the scale factors (`xscl` and `yscl`) we need to multiply the pixels by (in this case, 0 to 600) in order to get the complex numbers (in this case, between –2 and 2). In the `draw()` function we define our complex number `z`, and then we feed it into the `mandelbrot()` function and print out what we get. Nothing will appear on the screen yet, but in the console, you'll see the number 4 printed out. Now we'll go through every pixel on the screen and put their location into the `mandelbrot()` function and display the results.

Let's return to our `mandelbrot()` function in the *mandelbrot.pyde* program. Repeating the multiplication and addition operations on a pixel's location returns a number, and if the number never diverges, we color the pixel black. The entire `draw()` function is shown in Listing 7-5.

mandelbrot.pyde

```
def draw():
    #origin in center:
    translate(width/2,height/2)
    #go over all x's and y's on the grid
❶   for x in range(width):
        for y in range(height):
❷           z = [(xmin + x * xscl) ,
                 (ymin + y * yscl) ]
            #put it into the mandelbrot function
❸           col=mandelbrot(z,100)
            #if mandelbrot returns 0
            if col == 100:
                fill(0) #make the rectangle black
            else:
                fill(255) #make the rectangle white
            #draw a tiny rectangle
            rect(x,y,1,1)
```

Listing 7-5: Looping over all the pixels in the display window

Going over all the pixels requires a nested loop for x and y ❶. We declare complex number z to be x + *i*y ❷. Calculating the complex number z from the window coordinates is a little tricky. We start at the xmin value, for instance, and add the number of steps we're taking multiplied by the scale factor. We're not going between 0 and 600, which is the size of the display window in pixels; we're just going between –2 and 2. We run that through the mandelbrot() function ❸.

The mandelbrot() function squares and adds the complex number 100 times and returns the number of iterations it took for the number to diverge. This number is saved to a variable called col since color is already a keyword in Processing. The number in col determines what color we make that pixel. For now, let's just get a Mandelbrot set on the screen by making every pixel that never diverges black. Otherwise, we'll make the rectangle white. Run this code and you should see the famous Mandelbrot set, like in Figure 7-8.

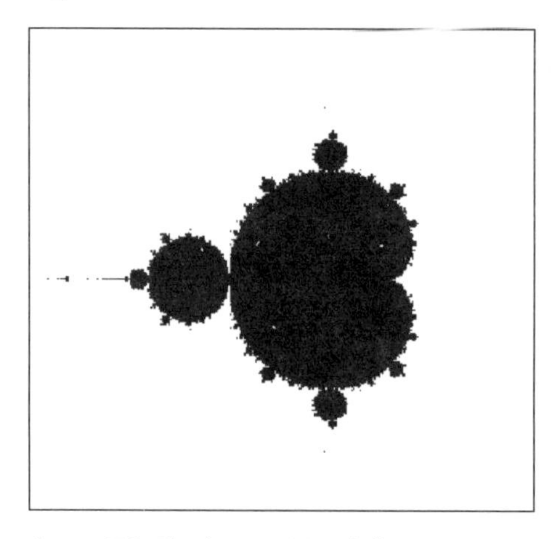

Figure 7-8: The famous Mandelbrot set

Isn't it amazing? And it's definitely unexpected: just by squaring and adding complex numbers, and coloring the pixels according to how large the numbers get, we've drawn an infinitely complicated design that could never have been imagined without a computer! You can zoom in on specific spots in the design by changing the range of x and y, like in Listing 7-6.

```
#range of x-values
xmin = -0.25
xmax = 0.25

#range of y-values
ymin = -1
ymax = -0.5
```

Listing 7-6: Changing the range of values to zoom in on the Mandelbrot set

Now the output should look like Figure 7-9.

Figure 7-9: Zooming in on the Mandelbrot set!

I highly recommend that you investigate videos people have posted on the internet of zooming in on the Mandelbrot set.

ADDING COLOR TO THE MANDELBROT SET

Now let's add some color to your Mandelbrot design. Let Processing know you're using the HSB (Hue, Saturation, Brightness) scale, not the RGB (Red, Green, Blue) scale, by adding the following code:

```
def setup():
    size(600,600)
    colorMode(HSB)
    noStroke()
```

Then color the rectangles according to the value returned by the `mandelbrot()` function:

```
            if col == 100:
                fill(0)
            else:
                fill(3*col,255,255)
            #draw a tiny rectangle
            rect(x*xscl,y*yscl,1,1)
```

In the `fill` line, we multiply the `col` variable (the number of iterations it takes the complex number to diverge) by 3 and make that the H (hue) component of the HSB color mode. Run this code, and you should see a nicely colored Mandelbrot set like in Figure 7-10.

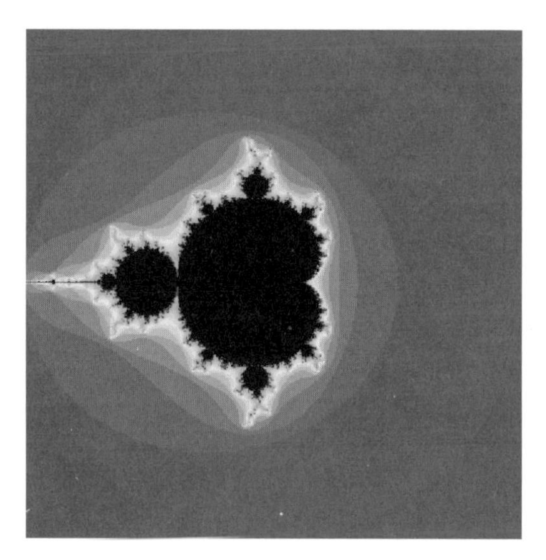

Figure 7-10: Using divergence values to color the Mandelbrot set

You can see the points that diverge every step, from the dark orange circle to lighter orange ovals that become the black Mandelbrot set. You can experiment with other colors too. For example, change the fill line to the following:

```
fill(255-15*col,255,255)
```

Run this update, and you'll see more blue in the picture, as shown in Figure 7-11.

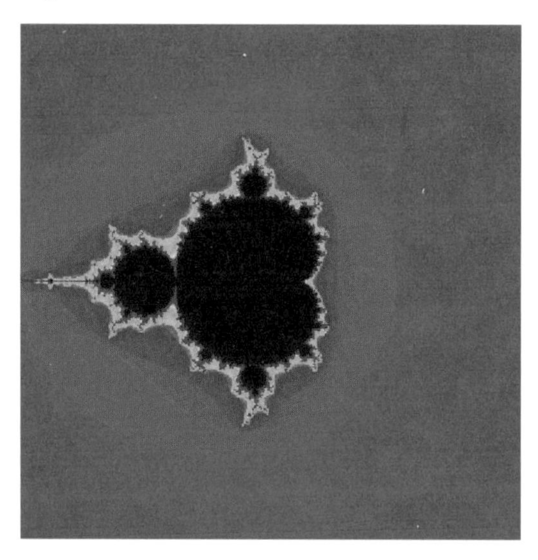

Figure 7-11: Experimenting with different colors in the Mandelbrot set

Next, we'll explore a related design called the Julia set, which can change its appearance depending on the inputs we give it.

CREATING THE JULIA SET

In the Mandelbrot set, to determine the color of each point, we started with the point as a complex number z and then repeatedly squared the value and added the original number z. The Julia set is constructed just like the Mandelbrot set, but after squaring the complex number, instead of adding the original complex number of that point, we keep adding a constant complex number, *c*, which has the same value for all points. By starting with different values for *c*, we can create lots of different Julia sets.

WRITING THE JULIA() FUNCTION

The Wikipedia page for the Julia set gives a bunch of examples of beautiful Julia sets and the complex numbers to use to create them. Let's try to create one using $c = -0.8 + 0.156i$. We can easily modify our `mandelbrot()` function to be a `julia()` function. Save your *mandelbrot.pyde* sketch as *julia.pyde* and change the code for the `mandelbrot()` function so it looks like Listing 7-7.

julia.pyde

```
def julia(z,c,num):
    '''runs the process num times
    and returns the diverge count'''
    count = 0
    #define z1 as z
    z1 = z
    #iterate num times
    while count <= num:
        #check for divergence
        if magnitude(z1) > 2.0:
            #return the step it diverged on
            return count
        #iterate z
      ❶ z1 = cAdd(cMult(z1,z1),c)
        count += 1
```

Listing 7-7: Writing the `julia()` function

It's pretty much the same as the Mandelbrot function. The only line of code that changed is ❶, where z is changed to c. The complex number `c` will be different from `z`, so we'll have to pass that to the `julia()` function in `draw()`, as shown in Listing 7-8.

```
def draw():
    #origin in center:
    translate(width/2,height/2)
    #go over all x's and y's on the grid
    x = xmin
    while x < xmax:
        y = ymin
```

```
while y < ymax:
    z = [x,y]
  ❶ c = [-0.8,0.156]
    #put it into the julia program
    col = julia(z,c,100)
    #if julia returns 100
    if col == 100:
        fill(0)
    else:
        #map the color from 0 to 100
        #to 0 to 255
        #coll = map(col,0,100,0,300)
        fill(3*col,255,255)
    rect(x*xscl,y*yscl,1,1)
    y += 0.01
x += 0.01
```

Listing 7-8: Writing the draw() function for the Julia set

Everything is the same as in *mandelbrot.pyde* until we declare the complex number c ❶ we've chosen for this Julia set. Just below that we add c to the arguments in the call to the `julia()` function. When you run it, you get a design much different from the Mandelbrot set, as shown in Figure 7-12.

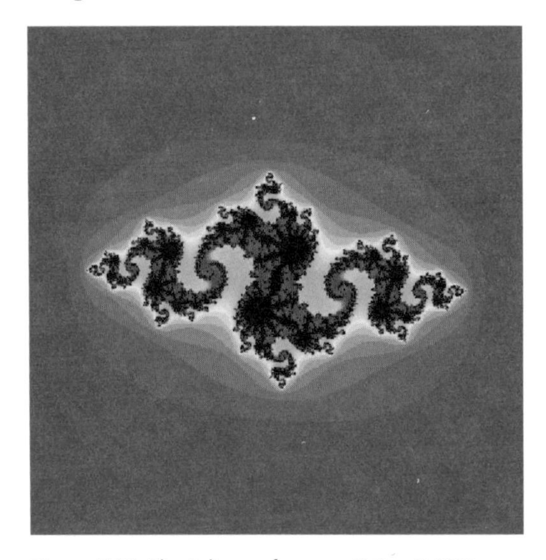

Figure 7-12: The Julia set for c = –0.8 + 0.156i

The great thing about the Julia set is you can change the input c and get a different output. For example, if you change c to $0.4 + 0.6i$, you should see something like Figure 7-13.

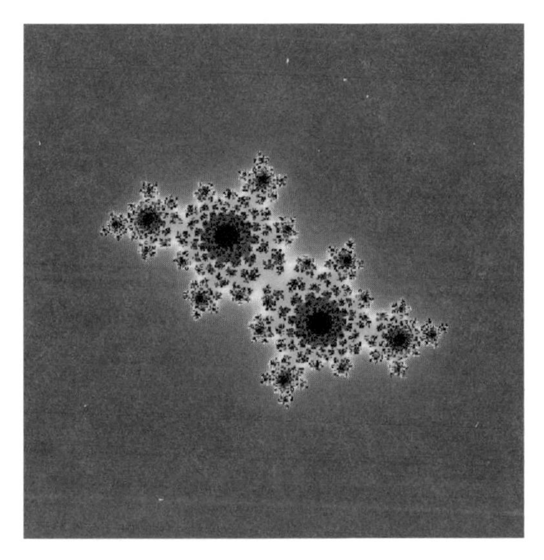

Figure 7-13: The Julia set for c = *−0.4 + 0.6*i

> **EXERCISE 7-1: DRAWING A JULIA SET**
>
> Draw a Julia set with $c = 0.285 + 0.01i$.

SUMMARY

In this chapter, you learned how complex numbers get plotted on the complex coordinate plane and how they allow you to perform rotations—and you followed their logic down the rabbit hole, learning how to add and multiply them. You used what you learned to write the `mandelbrot()` and `julia()` functions to transform complex numbers into incredible art that never would have been possible without the creation of complex numbers and the invention of computers.

As you've seen, these numbers are anything but imaginary! Hopefully, when you think of complex numbers now, they'll remind you of the beautiful designs you can make with numbers and code.

USING MATRICES FOR COMPUTER GRAPHICS AND SYSTEMS OF EQUATIONS

"I am large, I contain multitudes."
—Walt Whitman, from "Song of Myself"

In math class, students are taught how to add, subtract, and multiply matrices without ever learning how they're really used. This is too bad because matrices allow us to easily group together large collections of items and simulate coordinates of an object from multiple perspectives, making them useful in machine learning and absolutely crucial to 2D and 3D graphics. In other words, without matrices, there would be no video games!

To understand how matrices are useful for creating graphics, you first need to understand how to perform arithmetic on them. In this chapter, you review how to add and multiply matrices so that you can create and transform 2D and 3D objects in Processing. Finally, you learn how to solve large systems of equations instantaneously using matrices.

WHAT IS A MATRIX?

A *matrix* is a rectangular array of numbers that have specific rules for operating on them. Figure 8-1 shows what a matrix looks like.

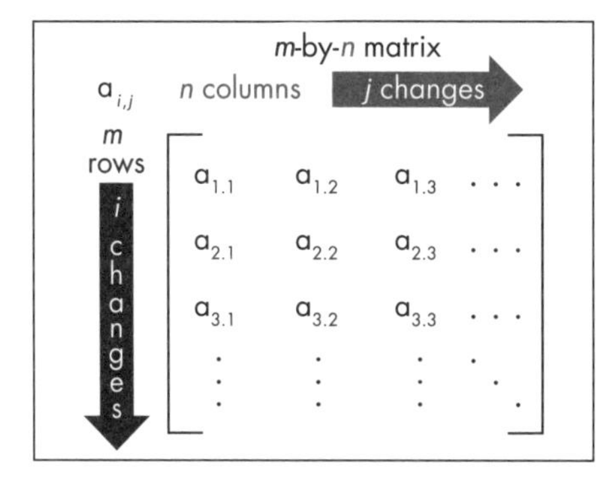

Figure 8-1: Matrices have m *rows and* n *columns*

Here, the numbers are arranged in rows and columns, where m and n represent the total number of rows and columns, respectively. You can have a 2 × 2 matrix, with two rows and two columns, like this:

$$\begin{bmatrix} 1 & 5 \\ -9 & 2 \end{bmatrix}$$

Or you can have a 3 × 4 matrix with three rows and four columns, like this:

$$\begin{bmatrix} 4 & -3 & 11 & -13 \\ 1 & 0 & 7 & 20 \\ -12 & 2 & 5 & 6 \end{bmatrix}$$

Traditionally, we use the letter i to represent the row number and the letter j to represent the column number. Note that the numbers in a matrix aren't being added to each other; they're just arranged together. This is similar to how we arrange coordinates using the format (x,y), but you don't operate on coordinates. For example, a point at (2,3) doesn't mean you add or multiply 2 and 3; they just sit next to each other and tell you where the point is located on a graph. But as you'll soon see, you *can* add, subtract, and multiply two matrices just like you can normal numbers.

ADDING MATRICES

You can only add and subtract matrices of the same dimensions (size and shape), which means that you can add or subtract only the *corresponding elements*. Here is an example of how to add two 2 × 2 matrices:

$$\begin{bmatrix} \textcircled{1} & -2 \\ \textcircled{3} & 4 \end{bmatrix} + \begin{bmatrix} \textcircled{5} & 6 \\ \textcircled{-7} & 8 \end{bmatrix} = \begin{bmatrix} 6 & 4 \\ -4 & 12 \end{bmatrix}$$

For example, we add 1 and 5 because they are corresponding elements in their matrices, meaning they're in the same spot: the first row, first column. Thus, we get 6 in the top-left corner. Adding the corresponding elements 3 and –7 gives us –4, as you can see in the bottom-left corner of the result.

That's easy enough to put into a Python function since you can save a matrix to a variable. Open a new file in IDLE and save it as *matrices.py*. Then write the code in Listing 8-1.

matrices.py

```
A = [[2,3],[5,-8]]
B = [[1,-4],[8,-6]]

def addMatrices(a,b):
    '''adds two 2x2 matrices together'''
    C = [[a[0][0]+b[0][0],a[0][1]+b[0][1]],
         [a[1][0]+b[1][0],a[1][1]+b[1][1]]]
    return C

C = addMatrices(A,B)
print(C)
```

Listing 8-1: Writing the matrices.py *program to add matrices*

Here, we declare a couple of 2 × 2 matrices, `A` and `B`, using Python's list syntax. For example, `A` is a list that contains two lists, each of which has two elements. We then declare a function called `addMatrices()`, which takes two matrices as arguments. Finally, we create another matrix, `C`, which adds each element in the first matrix to the corresponding element in the second.

When you run this, the output should be something like this:

```
[[3, -1], [13, -14]]
```

This shows the 2 × 2 matrix that results from adding matrices A and B:

$$\begin{bmatrix} 3 & -1 \\ 13 & -14 \end{bmatrix}$$

Now that you know how to add matrices, let's try multiplying them, which will let you transform coordinates.

MULTIPLYING MATRICES

Multiplying matrices is much more useful than adding them. For example, you can rotate a 2D or 3D shape by multiplying a matrix of (x,y) coordinates by a transformation matrix, as you'll do later in this chapter.

When multiplying matrices, you don't simply multiply the corresponding elements. Instead, you multiply the elements in each row of the first matrix by the corresponding elements in each column of the second matrix. This means that the number of columns in the first matrix has to equal the number of rows in the second. Otherwise, they can't be multiplied. For example, the following two matrices can be multiplied:

$$\begin{bmatrix} 1 & 2 \\ 3 & 4 \end{bmatrix} \begin{bmatrix} 5 \\ 6 \end{bmatrix}$$

First, we multiply the elements in the first row of the first matrix (1 and 2) with the elements in only the first column of the second matrix (5 and 6). The sum of those products would become the element in the first row and column of the resultant matrix. We do the same for the second row of the first matrix, and so on. It would look like this:

$$\begin{bmatrix} 1 & 2 \\ 3 & 4 \end{bmatrix} \begin{bmatrix} 5 \\ 6 \end{bmatrix} = \begin{bmatrix} 1 \times 5 + 2 \times 6 \\ 3 \times 5 + 4 \times 6 \end{bmatrix} = \begin{bmatrix} 17 \\ 39 \end{bmatrix}$$

Here is the general formula for multiplying a 2 × 2 matrix by a 2 × 2 matrix:

$$\begin{bmatrix} a & b \\ c & d \end{bmatrix} \begin{bmatrix} e & f \\ g & h \end{bmatrix} = \begin{bmatrix} ae + bg & af + bh \\ ce + dg & ch + dh \end{bmatrix}$$

We can also multiply the following two matrices, because A is a 1 × 4 matrix and B is a 4 × 2 matrix:

$$A = \begin{bmatrix} 1 & 2 & -3 & -1 \end{bmatrix}$$

$$B = \begin{bmatrix} 4 & -1 \\ -2 & 3 \\ 6 & -3 \\ 1 & 0 \end{bmatrix}$$

What will the resultant matrix look like? Well, the first row of A will be multiplied by the first column of B to become the number in the first row, first column of the result. It works the same way for the first row, second column. The resultant matrix will be a 1 × 2 matrix. You can see when you're multiplying matrices, the elements in the rows of the first matrix are being matched up with the elements in the columns of the second matrix. That means the resultant matrix will have the number of rows of the first matrix and the number of columns of the second matrix.

Now we'll directly multiply the elements in matrix A by their corresponding elements in matrix B and add all the products.

$$AB = \begin{bmatrix} 1 \times 4 + 2 \times -2 + -3 \times 6 + -1 \times 1 & 1 \times -1 + 2 \times 3 + -3 \times -3 + -1 \times 0 \end{bmatrix}$$

$$AB = \begin{bmatrix} -19 & 14 \end{bmatrix}$$

This might seem like a complicated process to have to automate, but as long as we have the matrices as input, we can easily find out the number of columns and rows.

Listing 8-2 shows a matrix multiplication program in Python that requires a bit more work than the addition code. Add this code to *matrices.py*.

```
def multmatrix(a,b):
    #Returns the product of matrix a and matrix b
    m = len(a) #number of rows in first matrix
    n = len(b[0]) #number of columns in second matrix
    newmatrix = []
    for i in range(m):
        row = []
        #for every column in b
        for j in range(n):
            sum1 = 0
            #for every element in the column
            for k in range(len(b)):
                sum1 += a[i][k]*b[k][j]
            row.append(sum1)
        newmatrix.append(row)
    return newmatrix
```

Listing 8-2: Writing a matrix multiplication function

In this example, the `multmatrix()` function takes two matrices as parameters: `a` and `b`. Right at the beginning of the function we declare `m`, the number of rows in matrix `a`, and `n`, the number of columns in matrix `b`. We create an empty list called `newmatrix` as the resultant matrix. The "row times column" operation will occur `m` times, so the first loop is `for i in range(m)`, making `i` repeat `m` number of times. For every row, we add an empty row to `newmatrix` so we can fill the row with `n` elements. The next loop makes `j` repeat `n` times because there are `n` columns in `b`. The tricky part will be matching up the correct elements, but it just takes a little thinking.

Just think of what elements will be multiplied together. When `j = 0`, we multiply the elements in the `i`th row of `a` by the first column (index 0) of `b`, and the product becomes the first column in the new row of `newmatrix`, as you saw in the previous example. Then, when `j = 1`, the same happens to the `i`th row of `a` and the second column (index 1) of `b`. That product becomes the second column in the new row of `newmatrix`. This process gets repeated for every row of `a`.

For every element in the row in matrix `a`, there's a corresponding element in the column in matrix `b`. The number of columns of `a` and the number of rows of `b` are the same, but we can express it as `len(a[0])` or `len(b)`. I chose `len(b)`. So in the third loop, `k` will repeat `len(b)` times. The first element in the `i`th row of `a` and the first element in the `j`th column of `b` will be multiplied together, which can be written like this:

```
a[i][0] * b[0][j]
```

The same for the second element in the `i`th row of `a` and the second element in the `j`th column of `b`:

```
a[i][1] * b[1][j]
```

So for every column (in the j loop), we'll start a running sum at 0 (because sum is already a Python keyword, I use sum1), and it will increment for every one of the k elements:

```
sum1 += a[i][k] * b[k][j]
```

It doesn't look like much, but that's the line that's going to keep track of and multiply all the corresponding elements! After going through all k elements (after the k loop is finished), we'll append the sum to the row, and once we've gone through all the columns in b (after the j loop is finished), we'll put that row into newmatrix. After going through all the rows in a, we return the resultant matrix.

Let's test this program by multiplying our sample matrices, a 1×4 matrix by a 4×2 matrix:

```
>>> a = [[1,2,-3,-1]]
>>> b = [[4,-1],
         [-2,3],
         [6,-3],
         [1,0]]
>>> print(multmatrix(a,b))
[[-19, 14]]
```

This checks out:

$$(1)(4) + (2)(-2) + (-3)(6) + (-1)(1) = -19$$

and

$$(1)(-1) + (2)(3) + (-3)(-3) + (-1)(0) = 14$$

Therefore, our new function for multiplying any two matrices (if they *can* be multiplied) works. Let's test it by multiplying a 2×2 matrix by a 2×2 matrix:

$$a = \begin{bmatrix} 1 & -2 \\ 2 & 1 \end{bmatrix}$$

$$b = \begin{bmatrix} 3 & -4 \\ 5 & 6 \end{bmatrix}$$

Enter the following to multiply matrix a by matrix b:

```
>>> a = [[1,-2],[2,1]]
>>> b = [[3,-4],[5,6]]
>>> multmatrix(a,b)
[[-7, -16], [11, -2]]
```

The code shows how to enter 2×2 matrices using Python lists. The multiplication also looks like this:

$$\begin{bmatrix} 1 & -2 \\ 2 & 1 \end{bmatrix} \begin{bmatrix} 3 & -4 \\ 5 & 6 \end{bmatrix} = \begin{bmatrix} -7 & -16 \\ 2 & -2 \end{bmatrix}$$

Let's check these answers. We begin by multiplying the first row of `a` by the first column of `b`:

$$(1)(3) + (-2)(5) = 3 - 10 = -7$$

And –7 is the number in the first row, first column of the resultant matrix. We next multiply the second row of `a` by the first column of `b`:

$$(2)(3) + (1)(5) = 6 + 5 = 11$$

And 11 is the number in the second row, first column of the resultant matrix. The other numbers are correct, too. The `multmatrix()` function is going to save us from doing a lot of laborious arithmetic!

ORDER MATTERS IN MATRIX MULTIPLICATION

An important fact about multiplying matrices is that A × B doesn't necessarily equal B × A. Let's prove that by reversing our previous example:

$$\begin{bmatrix} 3 & -4 \\ 5 & 6 \end{bmatrix} \begin{bmatrix} 1 & -2 \\ 2 & 1 \end{bmatrix} = \begin{bmatrix} -5 & -10 \\ 17 & -4 \end{bmatrix}$$

Here's how to multiply these matrices in the other direction in the Python shell:

```
>>> a = [[1,-2],[2,1]]
>>> b = [[3,-4],[5,6]]
>>> multmatrix(b,a)
[[-5, -10], [17, -4]]
```

As you can see, when you multiply the same matrices in the reverse order using `multmatrix(b,a)` instead of `multmatrix(a,b)`, you get a completely different resultant matrix. Remember that when you're multiplying matrices, *A × B doesn't necessarily equal B × A.*

DRAWING 2D SHAPES

Now that you know how to operate on matrices, let's put a bunch of points into a list to make a 2D shape. Open a new sketch in Processing and save it as *matrices.pyde*. If you still have your *grid.pyde* sketch from Listing 4-11, you can copy and paste the essentials for drawing a grid. Otherwise, add the code in Listing 8-3.

matrices.pyde

```
#set the range of x-values
xmin = -10
xmax = 10

#range of y-values
ymin = -10
ymax = 10
```

```
#calculate the range
rangex = xmax - xmin
rangey = ymax - ymin

def setup():
    global xscl, yscl
    size(600,600)
    #the scale factors for drawing on the grid:
    xscl= width/rangex
    yscl= -height/rangey
    noFill()

def draw():
    global xscl, yscl
    background(255) #white
    translate(width/2,height/2)
    grid(xscl, yscl)

def grid(xscl,yscl):
    '''Draws a grid for graphing'''
    #cyan lines
    strokeWeight(1)
    stroke(0,255,255)
    for i in range(xmin,xmax+1):
        line(i*xscl,ymin*yscl,i*xscl,ymax*yscl)
    for i in range(ymin,ymax+1):
        line(xmin*xscl,i*yscl,xmax*xscl,i*yscl)
    stroke(0) #black axes
    line(0,ymin*yscl,0,ymax*yscl)
    line(xmin*xscl,0,xmax*xscl,0)
```

Listing 8-3: The code for drawing a grid

We're going to draw a simple figure and transform it using matrices. I'll use the letter *F* because it has no rotational or reflectional symmetry (and because it's my initial). We'll sketch it out to get the points, as shown in Figure 8-2.

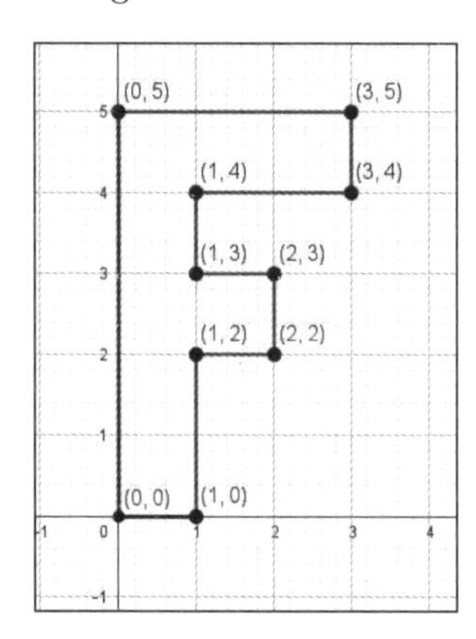

Figure 8-2: The points needed to draw an F

Add the code in Listing 8-4 after the draw() function to enter the points for all the corners of the *F* and draw lines between those points.

```
fmatrix = [[0,0],[1,0],[1,2],[2,2],[2,3],[1,3],[1,4],[3,4],[3,5],[0,5]]

def graphPoints(matrix):
    #draw line segments between consecutive points
    beginShape()
    for pt in matrix:
        vertex(pt[0]*xscl,pt[1]*yscl)
    endShape(CLOSE)
```

Listing 8-4: Graphing the points to draw the F

Here, we first create a list called fmatrix and enter points on each row corresponding to the points in the letter *F*. The graphPoints() function takes a matrix as a parameter, and each row becomes a vertex of the shape using Processing's beginShape() and endShape() functions. Also, we call the graphPoints() function using fmatrix as an argument in the draw() function. Add the code in Listing 8-5 to the end of the draw() function:

```
    strokeWeight(2) #thicker line
    stroke(0) #black
    graphPoints(fmatrix)
```

Listing 8-5: Getting the program to graph the points in the F

We're creating the fmatrix as a list containing a bunch of coordinates, and we call the graphPoints() function to tell the program to graph all the points.

Processing's built-in strokeWeight() function lets you control how thick you want the outline to be, and the stroke() function lets you choose the color of the outline. We'll draw the first *F* in black. The output looks like Figure 8-3.

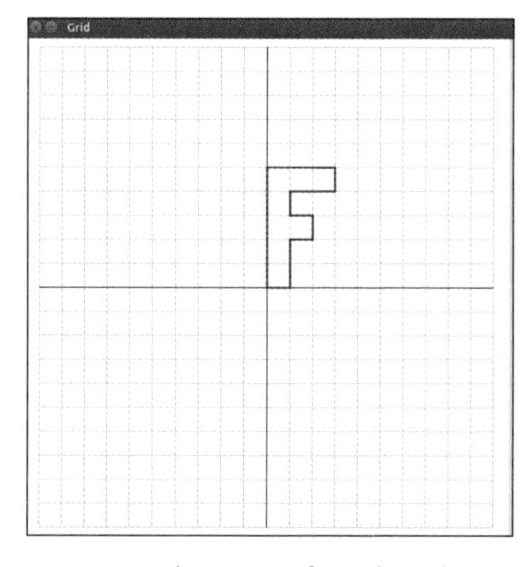

Figure 8-3: The output of graphing the points in the matrix, called the "f-matrix"

When we learn about matrices in school, we learn how to add and multiply them, but we never learn why. It's only when you graph them that you realize that multiplying matrices is *transforming* them. Next, we'll use matrix multiplication to transform our *F*.

TRANSFORMING MATRICES

To see how multiplying matrices lets you transform them, we'll use a 2 × 2 transformation matrix I found on the web (see Figure 8-4).

In $\mathbb{R}^2$, consider the matrix that rotates a given vector $\mathbf{v}_0$ by a counterclockwise angle θ in a fixed coordinate system. Then

$$R_\theta = \begin{bmatrix} \cos\theta & -\sin\theta \\ \sin\theta & \cos\theta \end{bmatrix},$$

so

$$\mathbf{v}' = R_\theta \mathbf{v}_0.$$

Figure 8-4: A transformation matrix found online at mathworld.wolfram.com

It's going to rotate our *F* counterclockwise by an angle, given by theta (θ). If the angle is 90 degrees, cos(90) = 0 and sin(90) = 1. Therefore, the rotation matrix for a counterclockwise rotation of 90 degrees is

$$R = \begin{bmatrix} 0 & -1 \\ 1 & 0 \end{bmatrix}$$

We can create a transformation matrix by adding the following code to *matrices.pyde* before the `setup()` function:

```
transformation_matrix = [[0,-1],[1,0]]
```

Next, we multiply the f-matrix by the transformation matrix and save the result to a new matrix. Since the f-matrix is a 10 × 2 matrix and the transformation matrix is 2 × 2, the only way to multiply them is F × T, not T × F.

Remember, the number of columns in the first matrix has to equal the number of rows in the second matrix. We'll graph the f-matrix in black and change the stroke color to red for the new matrix. Replace `graphPoints(fmatrix)` by adding the following code in Listing 8-6 to the `draw()` function.

```
    newmatrix = multmatrix(fmatrix,transformation_matrix)
    graphPoints(fmatrix)
    stroke(255,0,0) #red resultant matrix
    graphPoints(newmatrix)
```

Listing 8-6: Multiplying the matrices and graphing the points

When you run this, it will look like Figure 8-5.

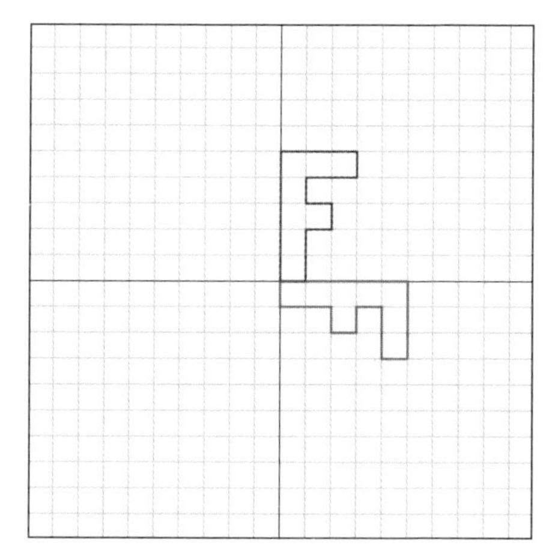

Figure 8-5: A clockwise rotation?

That's not a counterclockwise rotation! Looking at the math notation in Figure 8-4 again, we see the order of the multiplication is different from ours. The accepted way is to multiply by the transformation matrix first and then the point(s) to be transformed:

$$v' = R_\theta \, v_0$$

This means the transformed vector v (v') is the result of the rotation vector R_θ being multiplied by the initial vector v_0. Vector notation is different from coordinate notation. For instance, the vector that goes 2 in the x-direction and 3 in the y-direction is not given as (2,3), like in standard (x,y) coordinates. Rather, it's given as

$$\begin{bmatrix} 2 \\ 3 \end{bmatrix}$$

like a 2 × 1 matrix, instead of a 1 × 2 matrix. In our list notation, we'd write that as `[[2],[3]]`. That means we have to change our f-matrix to

```
fmatrix = [[[0],[0]],[[1],[0]],[[1],[2]],[[2],[2]],[[2],[3]],
           [[1],[3]],[[1],[4]],[[3],[4]],[[3],[5]],[[0],[5]]]
```

or

```
fmatrix = [[0,1,1,2,2,1,1,3,3,0],[0,0,2,2,3,3,4,4,5,5]]
```

The first example at least keeps the x- and y-values of a point together, but that's a lot of brackets! The second doesn't even keep the x- and y-values next to each other. Let's see if there's another way.

TRANSPOSING MATRICES

An important concept in matrices is *transposition*, where the columns become the rows, and vice versa. In our example, we want to change F into F^T, the notation for "the f-matrix, transposed."

$$F = \begin{bmatrix} 0 & 0 \\ 1 & 0 \\ 1 & 2 \\ 2 & 2 \\ 2 & 3 \\ 1 & 3 \\ 1 & 4 \\ 3 & 4 \\ 3 & 5 \\ 0 & 5 \end{bmatrix}$$

$$F^T = \begin{bmatrix} 0 & 1 & 1 & 2 & 2 & 1 & 1 & 3 & 3 & 0 \\ 0 & 0 & 2 & 2 & 3 & 3 & 4 & 4 & 5 & 5 \end{bmatrix}$$

Let's write a `transpose()` function that will transpose any matrix. Add the code in Listing 8-7 to *matrices.pyde* after the `draw()` function.

```
def transpose(a):
    '''Transposes matrix a'''
    output = []
    m = len(a)
    n = len(a[0])
    #create an n x m matrix
    for i in range(n):
        output.append([])
        for j in range(m):
            #replace a[i][j] with a[j][i]
            output[i].append(a[j][i])
    return output
```

Listing 8-7: The code to transpose a matrix

First, we create an empty list called `output` that will be the transposed matrix. We then define `m`, the number of rows in the matrix, and `n`, the number of columns. We're going to make output into an `n` × `m` matrix. For all `n` rows, we're going to start an empty list, and then everything in the `i`th row of the matrix we add to the `j`th column of the transposed matrix.

The following line of code in the `transpose` function switches the rows and columns of `a`:

```
output[i].append(a[j][i])
```

Finally, we return the transposed matrix. Let's test it out. Add the `transpose()` function to your *matrices.py* file and run it. Then we can enter the following code in the shell:

```
>>> a = [[1,2,-3,-1]]
>>> transpose(a)
[[1], [2], [-3], [-1]]
>>> b = [[4,-1],
        [-2,3],
        [6,-3],
        [1,0]]
>>> transpose(b)
[[4, -2, 6, 1], [-1, 3, -3, 0]]
```

It works! All we'll have to do is transpose our f-matrix before multiplying it by the transformation matrix. To graph it, we'll transpose it back, as shown in Listing 8-8.

matrices.pyde

```
def draw():
    global xscl, yscl
    background(255) #white
    translate(width/2,height/2)
    grid(xscl, yscl)
    strokeWeight(2) #thicker line
    stroke(0) #black
 ❶ newmatrix = transpose(multmatrix(transformation_matrix,
                         ❷ transpose(fmatrix)))
    graphPoints(fmatrix)
    stroke(255,0,0) #red resultant matrix
    graphPoints(newmatrix)
```

Listing 8-8: Transposing a matrix, multiplying, and then transposing again

Add the calls to the `transpose()` ❷ function to the `newmatrix` line ❶ of the `draw()` function. This should get you the correct counterclockwise rotation, as shown in Figure 8-6.

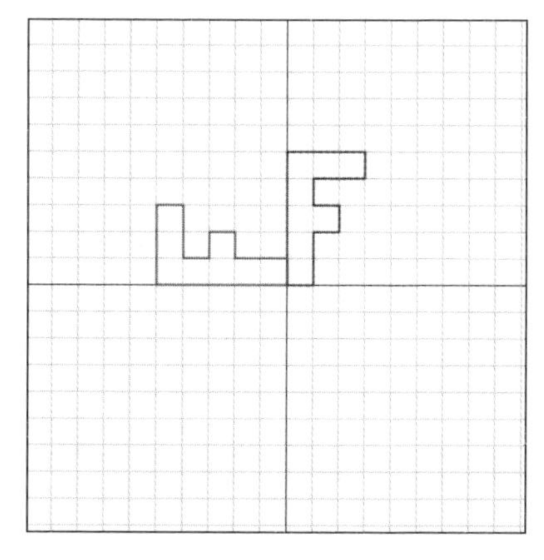

Figure 8-6: A counterclockwise rotation, by matrices

The final code for *matrices.pyde* should look like Listing 8-9.

matrices.pyde

```
#set the range of x-values
xmin = -10
xmax = 10

#range of y-values
ymin = -10
ymax = 10

#calculate the range
rangex = xmax - xmin
rangey = ymax - ymin

transformation_matrix = [[0,-1],[1,0]]

def setup():
    global xscl, yscl
    size(600,600)
    #the scale factors for drawing on the grid:
    xscl= width/rangex
    yscl= -height/rangey
    noFill()

def draw():
    global xscl, yscl
    background(255) #white
    translate(width/2,height/2)
    grid(xscl,yscl)
    strokeWeight(2) #thicker line
    stroke(0) #black
    newmatrix = transpose(multmatrix(transformation_matrix,
                          transpose(fmatrix)))
    graphPoints(fmatrix)
    stroke(255,0,0) #red resultant matrix
    graphPoints(newmatrix)

fmatrix = [[0,0],[1,0],[1,2],[2,2],[2,3],[1,3],[1,4],[3,4],[3,5],[0,5]]

def multmatrix(a,b):
    '''Returns the product of
    matrix a and matrix b'''
    m = len(a) #number of rows in first matrix
    n = len(b[0]) #number of columns in second matrix
    newmatrix = []
    for i in range(m): #for every row in a
        row = []
        #for every column in b
        for j in range(n):
            sum1 = 0
            #for every element in the column
```

```
            for k in range(len(b)):
                sum1 += a[i][k]*b[k][j]
            row.append(sum1)
        newmatrix.append(row)
    return newmatrix

def transpose(a):
    '''Transposes matrix a'''
    output = []
    m = len(a)
    n = len(a[0])
    #create an n x m matrix
    for i in range(n):
        output.append([])
        for j in range(m):
            #replace a[i][j] with a[j][i]
            output[i].append(a[j][i])
    return output

def graphPoints(matrix):
    #draw line segments between consecutive points
    beginShape()
    for pt in matrix:
        vertex(pt[0]*xscl,pt[1]*yscl)
    endShape(CLOSE)

def grid(xscl, yscl):
    '''Draws a grid for graphing'''
    #cyan lines
    strokeWeight(1)
    stroke(0,255,255)
    for i in range(xmin,xmax + 1):
        line(i*xscl,ymin*yscl,i*xscl,ymax*yscl)
    for i in range(ymin,ymax+1):
        line(xmin*xscl,i*yscl,xmax*xscl,i*yscl)
    stroke(0) #black axes
    line(0,ymin*yscl,0,ymax*yscl)
    line(xmin*xscl,0,xmax*xscl,0)
```

Listing 8-9: The entire code to draw and transform the letter F

EXERCISE 8-1: MORE TRANSFORMATION MATRICES

See what happens to your shape when you change your transformation matrix to each of these matrices:

a) $\begin{bmatrix} 1 & 0 \\ 0 & -1 \end{bmatrix}$ b) $\begin{bmatrix} 0 & -1 \\ -1 & 0 \end{bmatrix}$ c) $\begin{bmatrix} -1 & 1 \\ 1 & 1 \end{bmatrix}$

ROTATING MATRICES IN REAL TIME

So you just learned how matrices can transform points. But this can happen in real time, and interactively too! Change the code in the `draw()` function in *matrices.pyde* to what's in Listing 8-10.

```
def draw():
    global xscl, yscl
    background(255) #white
    translate(width/2,height/2)
    grid(xscl, yscl)
    ang = map(mouseX,0,width,0,TWO_PI)
    rot_matrix = [[cos(ang),-sin(ang)],
                  [sin(ang),cos(ang)]]
    newmatrix = transpose(multmatrix(rot_matrix,transpose(fmatrix)))
    graphPoints(fmatrix)
    strokeWeight(2) #thicker line
    stroke(255,0,0) #red resultant matrix
    graphPoints(newmatrix)
```

Listing 8-10: Rotating in real time using matrices

Recall that we used `sin()` and `cos()` in Chapter 7 to rotate and oscillate shapes. In this example, we're transforming a matrix of points using a rotation matrix. Here is what a typical 2 × 2 rotation matrix looks like:

$$R(\theta) = \begin{bmatrix} \cos(\theta) & -\sin(\theta) \\ \sin(\theta) & \cos(\theta) \end{bmatrix}$$

Because I don't have a theta (θ) key, I'll call the rotation angle `ang`. The interesting thing we're doing right now is changing the `ang` variable with the mouse. So, at every loop, the mouse position determines the value of `ang` and then plugs `ang` into each expression. It then quickly calculates the sine and cosine of `ang` and multiplies the rotation matrix by the f-matrix. For each loop, the rotation matrix will be a little different, depending on where your mouse is.

Now the red *F* should rotate around the origin as you move your mouse left and right over the graph, as shown in Figure 8-7.

This is the kind of transformation that's happening all the time when you see any kind of animation on a computer screen. Creating computer graphics is perhaps the most common application of matrices.

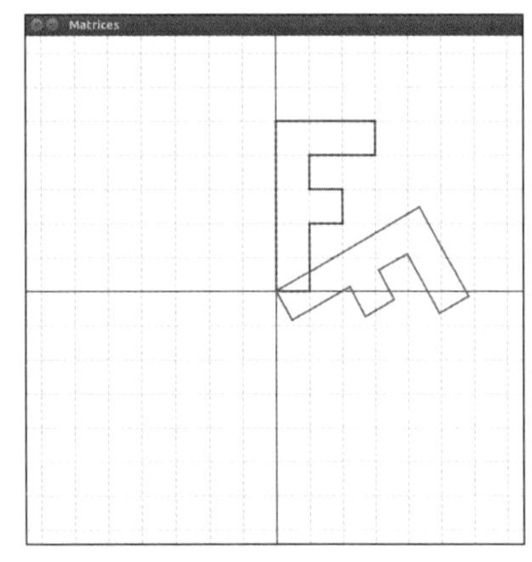

Figure 8-7: Transforming points in real time using matrices!

CREATING 3D SHAPES

So far we used matrices to create and manipulate two-dimensional shapes. You might be curious as to how we mathematicians crunch the numbers to represent a three-dimensional object on a two-dimensional surface like the computer screen.

Return to your code in Listing 8-11 and save it as *matrices3D.pyde*. Turn `fmatrix` into the following matrix of points:

```
fmatrix = [[0,0,0],[1,0,0],[1,2,0],[2,2,0],[2,3,0],[1,3,0],[1,4,0],
           [3,4,0],[3,5,0],[0,5,0],
           [0,0,1],[1,0,1],[1,2,1],[2,2,1],[2,3,1],[1,3,1],[1,4,1],
           [3,4,1],[3,5,1],[0,5,1]]
```

Listing 8-11: A 3D version of our f-matrix

Adding depth to our *F* requires adding another layer to our matrix of points. Because our *F* only has two dimensions right now, it's made up of only x- and y-values. But we can think of 2D objects as having a third dimension, represented by a z-axis. 2D objects have a z-value of 0. So for each point we'll add a zero as its third value, making the first 10 points three-dimensional. Then we'll copy and paste these values and change the third value to a 1. This creates the rear layer, which is an identical *F* drawn one unit behind the front one.

Now that we've created the two layers for the *F*, we need to connect the points on the front layer with those on the rear layer. Let's create an `edges` list so we can simply tell the program which points to link with line segments, as shown in Listing 8-12.

```
#list of points to connect:
edges = [[0,1],[1,2],[2,3],[3,4],[4,5],[5,6],[6,7],
         [7,8],[8,9],[9,0],
         [10,11],[11,12],[12,13],[13,14],[14,15],[15,16],[16,17],
         [17,18],[18,19],[19,10],
         [0,10],[1,11],[2,12],[3,13],[4,14],[5,15],[6,16],[7,17],
         [8,18],[9,19]]
```

Listing 8-12: Keeping track of the edges (the lines between the points on the F*)*

This is a way to keep track of which points are going to be connected with segments, or *edges*. For example, the first entry `[0,1]` draws an edge from point 0 (0,0,0) to point 1 (1,0,0). The first 10 edges draw the front *F*, and the next 10 edges draw the rear *F*. Then we draw edges between a point on the front *F* and the corresponding point on the rear *F*. For example, edge `[0,10]` draws a segment between point 0 (0,0,0) and point 10 (0,0,1).

Now when we're graphing the points, we're not just drawing lines between every consecutive point. Listing 8-13 shows the new `graphPoints()`

function that graphs the *edges* between the points in the list. Replace the old graphPoints() function with the following code, just before the definition of the grid() function.

```
def graphPoints(pointList,edges):
    '''Graphs the points in a list using segments'''
    for e in edges:
        line(pointList[e[0]][0]*xscl,pointList[e[0]][1]*yscl,
            pointList[e[1]][0]*xscl,pointList[e[1]][1]*yscl)
```

Listing 8-13: Graphing points using the edges

Remember that in Processing you draw a line between two points, (x1,y1) and (x2,y2), by using line(x1,y1,x2,y2). Here, we call the points in the pointList (we'll send fmatrix when we run this) by using their numbers in the edges list. The function loops over every item, e, in the edges list and connects the point represented by the first number, e[0], to the point represented by the second number, e[1]. The x-coordinates are multiplied by the xscl variable, which scales the x-values:

```
pointList[e[0]][0]*xscl
```

We do the same for the y-coordinates:

```
pointList[e[0]][1]*yscl
```

We can make our mouse represent the angle of rotation again by creating two rotation variables: rot and tilt. The first one, rot, maps the x-value of the mouse to an angle between 0 and 2π, and that value will go in a rotation matrix like the one we made in Listing 8-5. We do the same for tilt so it can map the y-value of the mouse. Put the code in Listing 8-14 in the draw() function before multiplying the matrices together.

```
    rot = map(mouseX,0,width,0,TWO_PI)
    tilt = map(mouseY,0,height,0,TWO_PI)
```

Listing 8-14: Linking the up-down and left-right rotations with the movements of the mouse

Next, we'll create a function to multiply the rotation matrices together so that all our transformations can be consolidated into one matrix. This is the great thing about using matrix multiplication to perform transformations. You can just keep "adding" more transformations simply by multiplying!

CREATING THE ROTATION MATRIX

Now let's make a single rotation matrix out of two individual matrices. If you see 3D rotation matrices in a math book, they may look like the following equations.

$$R_y(\theta) = \begin{bmatrix} \cos(\theta) & 0 & \sin(\theta) \\ 0 & 1 & 0 \\ -\sin(\theta) & 0 & \cos(\theta) \end{bmatrix}$$

$$R_x(\theta) = \begin{bmatrix} 1 & 0 & 0 \\ 0 & \cos(\theta) & \sin(\theta) \\ 0 & -\sin(\theta) & \cos(\theta) \end{bmatrix}$$

$R_y()$ will rotate the points, with the y-axis serving as the axis of rotation, so it's a left/right rotation. $R_x()$ will rotate the points around the x-axis, so it'll be an up-down rotation.

Listing 8-15 shows the code for creating the `rottilt()` function, which will take the `rot` and `tilt` values and put them into the matrices. This is how we combine two matrices into one. Add the code in Listing 8-15 to *matrices3D.pyde*:

```
def rottilt(rot,tilt):
    #returns the matrix for rotating a number of degrees
    rotmatrix_Y = [[cos(rot),0.0,sin(rot)],
                   [0.0,1.0,0.0],
                   [-sin(rot),0.0,cos(rot)]]
    rotmatrix_X = [[1.0,0.0,0.0],
                   [0.0,cos(tilt),sin(tilt)],
                   [0.0,-sin(tilt),cos(tilt)]]
    return multmatrix(rotmatrix_Y,rotmatrix_X)
```

Listing 8-15: Function for creating the rotation matrix

We multiply `rotmatrix_Y` and `rotmatrix_X` to get one rotation matrix as output. This is useful when there's a series of matrix operations, such as rotate about the x-axis R_x, rotate about the y-axis R_y, scale S, and translate T. Instead of performing a separate multiplication for each operation, we can combine all these operations in a single matrix. Matrix multiplication allows us to create a new matrix: $M = R_y(R_x(S(T)))$. That means our `draw()` function will change, too. With the new additions above, the `draw()` function should look like Listing 8-16:

```
def draw():
    global xscl, yscl
    background(255) #white
    translate(width/2,height/2)
    grid(xscl, yscl)
    rot = map(mouseX,0,width,0,TWO_PI)
    tilt = map(mouseY,0,height,0,TWO_PI)
    newmatrix = transpose(multmatrix(rottilt(rot,tilt),transpose(fmatrix)))
    strokeWeight(2) #thicker line
    stroke(255,0,0) #red resultant matrix
    graphPoints(newmatrix,edges)
```

Listing 8-16: The new draw() function

When you run the program, you get what's shown in Figure 8-8.

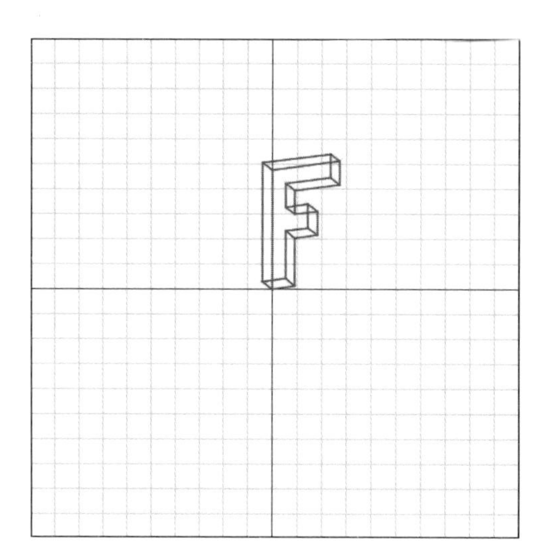

Figure 8-8: A 3D F!

We can remove the blue grid and make the *F* bigger by changing the `xmin`, `xmax`, `ymin`, and `ymax` variables and commenting out the call to the `grid()` function in `draw()`.

Listing 8-17 shows the full code for drawing a rotating 3D shape.

matrices3D.pyde

```
#set the range of x-values
xmin = -5
xmax = 5

#range of y-values
ymin = -5
ymax = 5

#calculate the range
rangex = xmax - xmin
rangey = ymax - ymin

def setup():
    global xscl, yscl
    size(600,600)
    #the scale factors for drawing on the grid:
    xscl= width/rangex
    yscl= -height/rangey
    noFill()

def draw():
    global xscl, yscl
    background(0) #black
    translate(width/2,height/2)
    rot = map(mouseX,0,width,0,TWO_PI)
    tilt = map(mouseY,0,height,0,TWO_PI)
    strokeWeight(2) #thicker line
    stroke(0) #black
    newmatrix = transpose(multmatrix(rottilt(rot,tilt),transpose(fmatrix)))
```

```
    #graphPoints(fmatrix)
    stroke(255,0,0) #red resultant matrix
    graphPoints(newmatrix,edges)

fmatrix = [[0,0,0],[1,0,0],[1,2,0],[2,2,0],[2,3,0],[1,3,0],[1,4,0],
          [3,4,0],[3,5,0],[0,5,0],
          [0,0,1],[1,0,1],[1,2,1],[2,2,1],[2,3,1],[1,3,1],[1,4,1],
          [3,4,1],[3,5,1],[0,5,1]]

#list of points to connect:
edges = [[0,1],[1,2],[2,3],[3,4],[4,5],[5,6],[6,7],
        [7,8],[8,9],[9,0],
        [10,11],[11,12],[12,13],[13,14],[14,15],[15,16],[16,17],
        [17,18],[18,19],[19,10],
        [0,10],[1,11],[2,12],[3,13],[4,14],[5,15],[6,16],[7,17],
        [8,18],[9,19]]

def rottilt(rot,tilt):
    #returns the matrix for rotating a number of degrees
    rotmatrix_Y = [[cos(rot),0.0,sin(rot)],
                   [0.0,1.0,0.0],
                   [-sin(rot),0.0,cos(rot)]]
    rotmatrix_X = [[1.0,0.0,0.0],
                   [0.0,cos(tilt),sin(tilt)],
                   [0.0,-sin(tilt),cos(tilt)]]
    return multmatrix(rotmatrix_Y,rotmatrix_X)

def multmatrix(a,b):
    '''Returns the product of
    matrix a and matrix b'''
    m = len(a) #number of rows in first matrix
    n = len(b[0]) #number of columns in second matrix
    newmatrix = []
    for i in range(m): #for every row in a
        row = []
        #for every column in b
        for j in range(n):
            sum1 = 0
            #for every element in the column
            for k in range(len(b)):
                sum1 += a[i][k]*b[k][j]
            row.append(sum1)
        newmatrix.append(row)
    return newmatrix

def graphPoints(pointList,edges):
    '''Graphs the points in a list using segments'''
    for e in edges:
        line(pointList[e[0]][0]*xscl,pointList[e[0]][1]*yscl,
             pointList[e[1]][0]*xscl,pointList[e[1]][1]*yscl)

def transpose(a):
    '''Transposes matrix a'''
    output = []
    m = len(a)
```

```
    n = len(a[0])
    #create an n x m matrix
    for i in range(n):
        output.append([])
        for j in range(m):
            #replace a[i][j] with a[j][i]
            output[i].append(a[j][i])
    return output
```

Listing 8-17: The full code for rotating the 3D F

I got rid of the grid and changed the call to the `background()` function in `draw()` to `background(0)`, so the background will be black and the *F* will appear to be rotating in outer space (see Figure 8-9)!

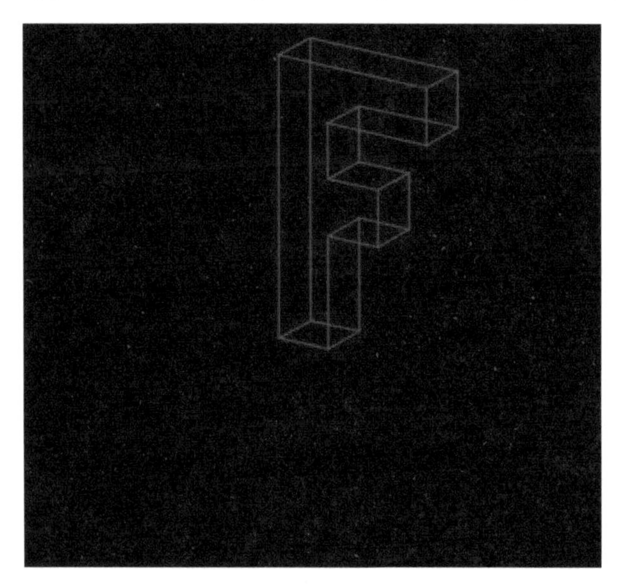

Figure 8-9: Moving your mouse around will transform the F!

SOLVING SYSTEMS OF EQUATIONS WITH MATRICES

Have you ever tried solving a system of equations with two or three unknown values? That's a tricky task for any math student. And as the number of unknowns increases, the more complicated the system of equations gets to solve. Matrices are very useful for solving systems of equations like this one:

$$2x + 5y = -1$$

$$3x - 4y = -13$$

For example, you can express this multiplication using matrices:

$$\begin{bmatrix} 2 & 5 \\ 3 & -4 \end{bmatrix} \begin{bmatrix} x \\ y \end{bmatrix} = \begin{bmatrix} -1 \\ -13 \end{bmatrix}$$

This looks similar to the algebra equation $2x = 10$, which we can solve easily by dividing both sides by 2. If only we could divide both sides of our system by the matrix on the left!

In fact, there is a way to do just that by finding the inverse of a matrix, the same way you can divide a number by 2 by multiplying it by ½. This is known as the *multiplicative inverse* of 2, but it's a complicated method.

GAUSSIAN ELIMINATION

The more efficient way to solve a system of equations using matrices is to use row operations to transform the 2 × 2 matrix on the left into the *identity matrix*, which is the matrix that represents the number 1. For example, multiplying a matrix by the identity matrix would result in the original matrix, like this:

$$\begin{bmatrix} 1 & 0 \\ 0 & 1 \end{bmatrix} \begin{bmatrix} x \\ y \end{bmatrix} = \begin{bmatrix} x \\ y \end{bmatrix}$$

The numbers on the right would be the solutions for x and y, so getting those zeroes and ones in the right place is our goal. The right place is the diagonal of the matrix, like this:

$$\begin{bmatrix} 1 & 0 \\ 0 & 1 \end{bmatrix} \quad or \quad \begin{bmatrix} 1 & 0 & 0 \\ 0 & 1 & 0 \\ 0 & 0 & 1 \end{bmatrix}$$

The identity matrix in every square matrix has a 1 on the diagonal, where the row number equals the column number.

Gaussian elimination is a method that involves doing operations on entire rows of matrices in order to get to the identity matrix. You can multiply or divide a row by a constant, and you can add or subtract a row from another row.

Before using Gaussian elimination, we first have to arrange the coefficients and constants into one matrix, like this:

$$\begin{bmatrix} 2 & 5 \\ 3 & -4 \end{bmatrix} \begin{bmatrix} x \\ y \end{bmatrix} = \begin{bmatrix} -1 \\ -13 \end{bmatrix} \rightarrow \begin{bmatrix} 2 & 5 & -1 \\ 3 & -4 & -13 \end{bmatrix}$$

Then, we divide the entire row by the number that will give us a 1 in the top left. This means that first we need to divide all the terms in the first row by 2, since 2/2 is 1. That operation gives us the following:

$$\begin{bmatrix} 1 & 5/2 & -1/2 \\ 3 & -4 & -13 \end{bmatrix}$$

Now we get the *additive inverse* (the number that gives us 0 when added to another number) of the term where we want a zero. For example, in row 2, we want a zero where the 3 is because we're looking to change this matrix into the identity matrix. Because the additive inverse of 3 is –3, we multiply each term in the first row by –3 and add the product to the corresponding term in row 2. That means we multiply the 1 in the first row by –3 and then add the product, which is still –3, to the second row. We repeat the process with all the terms in the row. For example, the –1/2 in the third column would be multiplied by –3 (to get 1.5) and added to all the numbers in that column. In this case, it's just –13, so the sum is –11.5 or –23/2. Continue this, and you should get the following:

$$\begin{bmatrix} 1 & 5/2 & -1/2 \\ 0 & -23/2 & -23/2 \end{bmatrix}$$

Now repeat where we want the 1 to be in the second row. We can multiply everything in row 2 by –2/23, which should give us this:

$$\begin{bmatrix} 1 & 5/2 & -1/2 \\ 0 & 1 & 1 \end{bmatrix}$$

Finally, we add everything in the first row to the second row, multiplied by the additive inverse of 5/2, which is where we want the zero to be in the first row. We'll add every term in the first row to its corresponding term in the second row multiplied by –5/2. Notice this doesn't affect the 1 in the first row, which we want to keep:

$$\begin{bmatrix} 1 & 0 & -3 \\ 0 & 1 & 1 \end{bmatrix}$$

The solutions to the system of equations are now in the right column: $x = -3$, $y = 1$.

We can check our answers by plugging those numbers into the original system of equations:

$$2(-3) + 5(1) = -6 + 5 = -1$$

$$3(-3) - 4(1) = -9 - 4 = -13$$

Both solutions are correct, but this is a laborious process. Let's automate this process with Python so we can solve systems that are as big as we want!

WRITING THE GAUSS() FUNCTION

In this section, we write a function called `gauss()` that solves systems of equations for us. Trying to do this programmatically might seem complicated, but there are really only two steps that we need to code:

1. Divide all the elements in a row by the term in the diagonal.
2. Add each term in one row to the corresponding term in another row.

Dividing All Items in a Row

The first task is to divide all the terms in a row by a number. Let's say we have the row of numbers `[1,2,3,4,5]`. For example, we can use the code in Listing 8-18 to divide this row by 2. Open a new Python file, call it *gauss.py* and enter the code in Listing 8-18.

```
divisor = 2
row = [1,2,3,4,5]
for i, term in enumerate(row):
    row[i] = term / divisor
print(row)
```

Listing 8-18: Dividing all the terms in a row by a divisor

This loops over the `row` list, keeping track of the index and the value using the `enumerate()` function. We then replace each term `row[i]` with the term divided by the divisor. When you run this, you'll get a list of five values:

```
[0.5, 1.0, 1.5, 2.0, 2.5]
```

Adding Each Element to Its Corresponding Element

The second task is to add each element of one row to the corresponding element in the other row. For example, add all the elements in row 0 below to the elements in row 1 and replace the elements in row 1 with the sum:

```
>>> my_matrix = [[2,-4,6,-8],
                 [-3,6,-9,12]]
>>> for i in range(len(my_matrix[1])):
        my_matrix[1][i] += my_matrix[0][i]
>>> print(my_matrix)
[[2, -4, 6, -8], [-1, 2, -3, 4]]
```

We're looping over all the items in the second row (index 1) of `my_matrix`. Then we're incrementing each term (index `i`) in the second row by the corresponding term in the first row (index 0). We successfully added the terms in the first row to those in the second row. Notice the first row didn't change. We'll use these steps in solving systems of equations.

Repeating the Process for Every Row

Now we just have to put those steps together for all the rows in a matrix. We'll call the matrix A. Once we've put the x, y, and z's and the constant terms in order, we put only the coefficients and the constants into the matrix:

$$A = \begin{bmatrix} 2 & 1 & -1 & 8 \\ -3 & -1 & 2 & -1 \\ -2 & 1 & 2 & -3 \end{bmatrix} \Leftarrow \begin{matrix} 2x + y - z = 8 \\ -3x - y + 2z = -1 \\ -2x + y + 2z = -3 \end{matrix}$$

First, we divide every term in the row by the term on the diagonal so that the diagonal term will be 1, using the code in Listing 8-19.

```
    for j,row in enumerate(A):
        #diagonal term to be 1
        #by dividing row by diagonal term
        if row[j] != 0: #diagonal term can't be 0
            divisor = row[j] #diagonal term
```

```
        for i, term in enumerate(row):
            row[i] = term / divisor
```

Listing 8-19: Dividing every term in a row by the row's diagonal term

Using enumerate, we can get the first row of A ([2,1,-1,8]), for example, and j will be the index of that row (in this case, zero). The diagonal terms are where the row number is the same as the column number, like row 0, column 0, or row 1, column 1.

Now we go through every other row in the matrix and perform the second step. Now, for each of the other rows (where i is not equal to j), calculate the additive inverse of the jth term, multiply each term in row j by that inverse, and add those terms to their corresponding terms in the ith row. Add the code in Listing 8-20 to the gauss() function.

```
        for i in range(m):
            if i != j: #don't do this to row j
                #calculate the additive inverse
                addinv = -1*A[i][j]
          #for every term in the ith row
          for ind in range(n):
              #add the corresponding term in the jth row
              #multiplied by the additive inverse
              #to the term in the ith row
              A[i][ind] += addinv*A[j][ind]
```

Listing 8-20: Making every nondiagonal term in a row 0

This happens to every row, so since m is the number of rows, we start off with for i in range(m). We already divided the row in question by the diagonal term, so we don't have to do anything to that row. That's why we do it only if i is not equal to j. In our example, each term in the first row of A is going to be multiplied by 3 and added to the corresponding term in the second row. And then every term in the first row is going to be multiplied by 2 and added to the corresponding term in the third row. That will get us zeroes in the second and third rows of the first column:

$$\begin{bmatrix} 1 & 1/2 & -1/2 & 4 \\ -3 & -1 & 2 & -1 \\ -2 & 1 & 2 & -3 \end{bmatrix} \longrightarrow \begin{bmatrix} 1 & 1/2 & -1/2 & 4 \\ 0 & 1/2 & 1/2 & 1 \\ 0 & 2 & 1 & 5 \end{bmatrix}$$

Now our first column is done, and we want a 1 in the diagonal. Therefore, we want a 1 in the second row of the second column, so we repeat the process.

Putting It All Together

Put all the code together into a gauss() function and print out the results. Listing 8-21 shows the complete code.

```
def gauss(A):
    '''Converts a matrix into the identity
    matrix by Gaussian elimination, with
```

```
    the last column containing the solutions
    for the variables'''
    m = len(A)
    n = len(A[0])
    for j,row in enumerate(A):
        #diagonal term to be 1
        #by dividing row by diagonal term
        if row[j] != 0: #diagonal entry can't be zero
            divisor = row[j]
            for i, term in enumerate(row):
                row[i] = term / divisor
        #add the other rows to the additive inverse
        #for every row
        for i in range(m):
            if i != j: #don't do it to row j
                #calculate the additive inverse
                addinv = -1*A[i][j]
                #for every term in the ith row
                for ind in range(n):
                    #add the corresponding term in the jth row
                    #multiplied by the additive inverse
                    #to the term in the ith row
                    A[i][ind] += addinv*A[j][ind]
    return A
#example:
B = [[2,1,-1,8],
     [-3,-1,2,-1],
     [-2,1,2,-3]]
print(gauss(B))
```

Listing 8-21: The complete code for the gauss() function

The output should be the following:

```
[[1.0, 0.0, 0.0, 32.0], [0.0, 1.0, 0.0, -17.0], [-0.0, -0.0, 1.0, 39.0]]
```

And here is how it looks in matrix form:

$$\begin{bmatrix} 1 & 0 & 0 & 32 \\ 0 & 1 & 0 & -17 \\ 0 & 0 & 1 & 39 \end{bmatrix}$$

We look at the last numbers in each row, so our solutions are $x = 32$, $y = -17$, and $z = 39$. We check this by plugging those values into the original equations:

$$2(32) + (-17) - (39) = 8. \text{ Check!}$$

$$-3(32) - (-17) + 2(39) = -1. \text{ Check!}$$

$$-2(32) + (-17) + 2(39) = -3. \text{ Check!}$$

This is a major achievement! Now, not only can we solve systems of two or three unknowns, but we can also solve for any number of unknowns! Solving a system of four unknowns is a laborious task if the student doesn't know Python. But luckily for us, we do! When the correct solution pops up so quickly in the Python shell, I'm always blown away. If you've ever had to perform Gaussian elimination by hand, Exercise 8-2 will blow you away too.

> **EXERCISE 8-2: ENTER THE MATRIX**
>
> Solve this system of equations for w, x, y, and z using the program you just wrote:
>
> $$2w - x + 5y + z = -3$$
> $$3w + 2x + 2y - 6z = -32$$
> $$w + 3x + 3y - z = -47$$
> $$5w - 2x - 3y + 3z = 49$$

SUMMARY

You've come a long way in your math adventure! You started with some basic Python to make turtles walk around and then went on to create more complicated Python functions to solve harder math problems. In this chapter, not only did you learn how to use Python to add and multiply matrices, but you also experienced first-hand how matrices can create an transform 2D and 3D graphics! The power we have to add, multiply, transpose, and otherwise operate on matrices using Python is mind-boggling.

You also learned to automate the process you would have done by hand to solve a system of equations. The same program that works for a 3 × 3 matrix will also work for a 4 × 4 or any larger square matrix!

Matrices are vital tools in making a neural network, with dozens or even hundreds of paths leading to and from virtual neurons. An input is "propagated" through the network using matrix multiplication and transposition, the same tools you created in this chapter.

There was a time when what you've done in this chapter was out of reach to anyone who didn't have access to an enormous, expensive computer that took up a whole floor at a university or major corporation. Now you can perform lightning-fast matrix calculations using Python and visualize the results using Processing!

In this chapter, I pointed out how great it is that we get instantaneous solutions to our complicated systems of equations as well as immediate responses to our mouse movements in our exploration of graphics. In the next chapter, we'll create a model of an ecosystem containing grass and sheep and let it run on its own. The model will morph and change over time as the sheep are born, eat, reproduce, and die. Only after letting the model run for a minute or more will we be able to judge if the environment can find a balance between grass growing and sheep eating and multiplying.

PART 3

BLAZING YOUR OWN TRAIL

BUILDING OBJECTS WITH CLASSES

Old teachers never die, they just lose their class.
—Anonymous

Now that you've created cool graphics using functions and other code in Processing, you can supercharge your creativity using classes. A *class* is a structure that lets you create new types of objects. The object types (usually just called *objects*) can have *properties,* which are variables, and *methods,* which are functions. There are times you want to draw multiple objects using Python, but drawing lots of them would be way too much work. Classes make drawing several objects with the same properties easy, but they require a specific syntax you'll need to learn.

The following example from the official Python website shows how to create a "Dog" object using a class. To code along, open a new file in IDLE, name it *dog.py* and enter the following code.

dog.py

```
class Dog:
    def __init__(self,name):
        self.name = name
```

This creates a new object called Dog using class Dog. It's customary in Python and many other languages to capitalize the name of a class, but it'll still work if you don't. To instantiate, or create, the class, we have to use Python's `__init__` method, which has two underscores before and two after init, meaning it's a special method to create (or *construct*) an object. The `__init__` line makes it possible to create instances of the class (in this case, dogs). In the `__init__` method, we can create any properties of the class we want. Since it's a dog, it can have a name, and because every dog has its own name, we use the self syntax. We don't need to use it when we call the objects, only when we're defining them.

We can then create a dog with a name using the following line of code:

```
d = Dog('Fido')
```

Now d is a Dog and its name is Fido. You can confirm this by running the file and entering the following in the shell:

```
>>> d.name
'Fido'
```

Now when we call d.name, we get Fido because that is the name property we just gave to it. We can create another Dog and give it the name Bettisa, like so:

```
>>> b = Dog('Bettisa')
>>> b.name
'Bettisa'
```

You can see one dog's name can be different from another's, but the program remembers them perfectly! This will be crucial when we give locations and other properties to the objects we create.

Finally, we can give the dog something to do by putting a function in the class. But don't call it a function! A function inside a class is called a *method*. Dogs bark, so we'll add that method to the code in Listing 9-1.

dog.py

```
class Dog:
    def __init__(self,name):
        self.name = name

    def bark(self):
        print("Woof!")

d = Dog('Fido')
```

Listing 9-1: Creating a dog that barks!

When we call the bark() method of the d dog, it barks:

```
>>> d.bark()
Woof!
```

It might not be clear why you'd need a Dog class from this simple example, but it's good to know you can do literally anything you want with classes and be as creative as you want. In this chapter, we use classes to make lots of useful objects like bouncing balls and grazing sheep. Let's start with the Bouncing Ball example to see how using classes lets us do something really cool while saving us a lot of work.

BOUNCING BALL PROGRAM

Start a Processing sketch and save it as *BouncingBall.pyde*. We'll draw a single circle on the screen, which we'll make into a bouncing ball. Listing 9-2 shows the code for drawing one circle.

BouncingBall.pyde

```
def setup():
    size(600,600)

def draw():
    background(0) #black
    ellipse(300,300,20,20)
```

Listing 9-2: Drawing a circle

First, we set the size of the window to be 600 pixels wide and 600 pixels tall. Then we set the background to black and drew a circle using the `ellipse()` function. The first two numbers in the function describe how far the center of the circle is from the top-left corner of the window, and the last two numbers describe the width and height of the ellipse. In this case, `ellipse(300,300, 20,20)` creates a circle that is 20 pixels wide and 20 pixels high, located in the center of the display window, as shown in Figure 9-1.

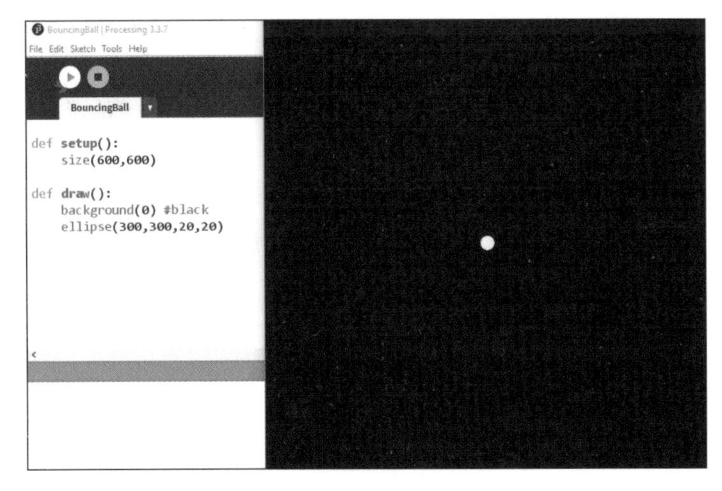

Figure 9-1: Drawing one circle for the Bouncing Ball sketch

Now that we have successfully created a circle located in the center of the display window, let's try to make it move.

MAKING THE BALL MOVE

We'll make the ball move by changing its position. To do this, let's first create a variable for the x-value and a variable for the y-value and set them to 300, which is the middle of the screen. Go back to Listing 9-2 and insert the following two lines at the beginning of the code, like in Listing 9-3.

BouncingBall.pyde

```
xcor = 300
ycor = 300

def setup():
    size(600,600)
```

Listing 9-3: Setting variables for the x- and y-values

Here, we use the `xcor` variable to represent the x-value and the `ycor` variable to represent the y-value. Then we set both variables to 300.

Now let's change the x-value and y-value by a certain number in order to change the location of the ellipse. Make sure to use the variables to draw the ellipse, as shown in Listing 9-4.

BouncingBall.pyde

```
xcor = 300
ycor = 300

def setup():
    size(600,600)

def draw():
❶   global xcor, ycor
    background(0) #black
    xcor += 1
    ycor += 1
    ellipse(xcor,ycor,20,20)
```

Listing 9-4: Incrementing `xcor` and `ycor` to change the location of the ellipse

The important thing to notice in this example is `global xcor, ycor` ❶, which tells Python to use the variables we've already created and not to create new ones just for the `draw()` function. If you don't include this line, you'll get an error message, something like "local variable 'xcor' referenced before assignment." Once Processing knows what value to assign to `xcor` and `ycor`, we increment them both by 1 and draw the ellipse with its center at the location specified using the global variables: `(xcor, ycor)`.

When you save and run Listing 9-4, you should see the ball move, like in Figure 9-2.

Now the ball moves down and to the right, because its x- and y-values are both increasing, but then it moves off the screen and we never see it again! The program keeps incrementing our variables obediently. It doesn't know it's drawing a ball or that we want the ball to bounce off the walls. Let's explore how to keep the ball from disappearing.

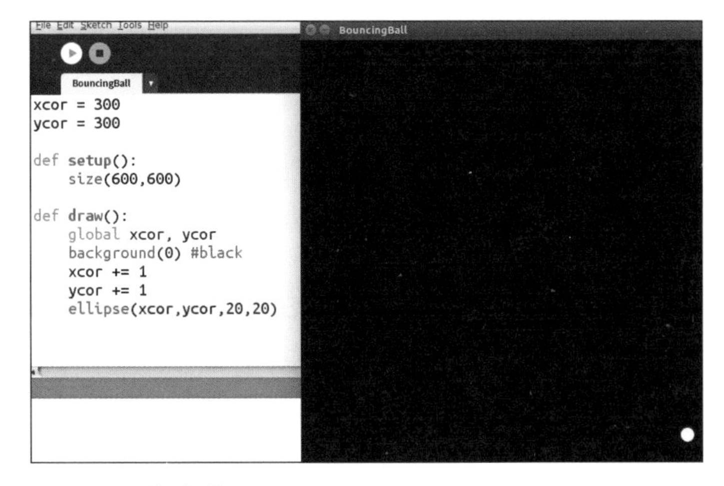

Figure 9-2: The ball moves!

MAKING THE BALL BOUNCE OFF THE WALL

When we change the x-value and the y-value by adding 1, we're changing the position of an object. In math, this change in position over time is called *velocity*. A positive change in x over time (positive x-velocity) will look like movement to the right (since x is getting bigger), whereas negative x-velocity will look like movement to the left. We can use this "positive-right, negative-left" concept to make the ball bounce off the wall. First, let's create the x-velocity and y-velocity variables by adding the following lines to our existing code, as shown in Listing 9-5.

BouncingBall.pyde

```
xcor = 300
ycor = 300
xvel = 1
yvel = 2

def setup():
    size(600,600)

def draw():
    global xcor,ycor,xvel,yvel
    background(0) #black
    xcor += xvel
    ycor += yvel
    #if the ball reaches a wall, switch direction.
    if xcor > width or xcor < 0:
        xvel = -xvel
    if ycor > height or ycor < 0:
        yvel = -yvel
    ellipse(xcor,ycor,20,20)
```

Listing 9-5: Adding code to make the ball bounce off the wall

First, we set `xvel = 1` and `yvel = 2` to specify how the ball will move. You can use other values and see how they change the movement. Then in the `draw()` function, we tell Python that `xvel` and `yvel` are global variables, and we change the x- and y-coordinates by incrementing using these variables. For example, when we set `xcor += xvel`, we're updating the position by the velocity (the *change* in position).

The two `if` statements tell the program that if the ball's position goes outside the boundaries of the screen, it should change the ball's velocity to its negative value. When we change the ball's velocity to its negative value, we tell the program to move the ball in the opposite direction it was moving in, making it seem like the ball is bouncing off the wall.

We need to be precise in telling at what point the ball should move in the opposite direction in terms of its coordinates. For example, `xcor > width` represents cases where `xcor` is larger than the width of the display window, which is when the ball touches the right edge of the screen. And `xcor < 0` represents instances where the `xcor` is less than 0 or when the ball touches the left edge of the screen. Similarly, `ycor > height` checks for instances where `ycor` is larger than the height of the window or when the ball reaches the bottom of the screen. Finally, `ycor < 0` checks for instances where the ball reaches the upper edge of the screen. Since moving to the right is positive x-velocity (positive change in x), the opposite direction is negative x-velocity. If the velocity is already negative (it's moving to the left), then the negative of a negative is a positive, which means the ball will move to the right, just like we want it to.

When you run Listing 9-5, you should see something like what's shown in Figure 9-3.

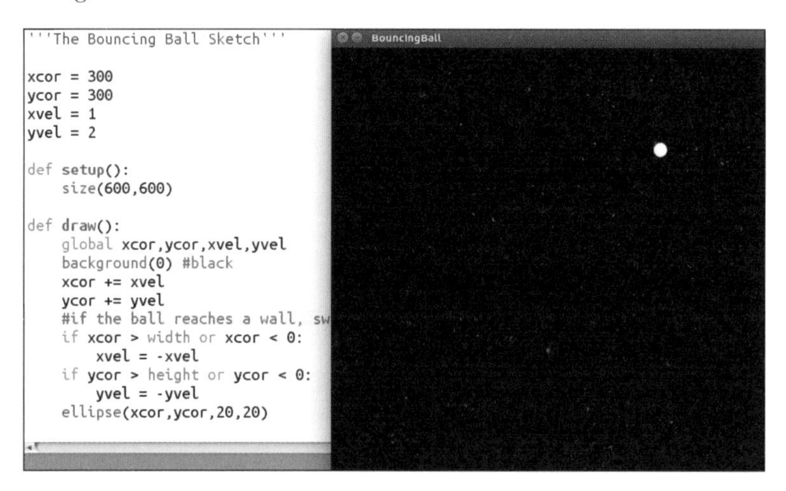

Figure 9-3: One bouncing ball!

The ball looks like it's bouncing off the walls and therefore stays in view.

MAKING MULTIPLE BALLS WITHOUT CLASSES

Now suppose we want to make another bouncing ball, or many other bouncing balls. How would we do that? We could make a new variable for the second ball's x-value, another variable for the second ball's y-value, a third variable for its x-velocity, and a fourth for its y-velocity. Then we'd have to increment its position by its velocity, check if it needs to bounce off a wall, and finally draw it. However, we'd end up with double the amount of code! Adding a third ball would triple our code! Twenty balls would simply be out of the question. You *don't* want to have to keep track of all these variables for position and velocity. Listing 9-6 show what this would look like.

```
#ball1:
ball1x = random(width)
ball1y = random(height)
ball1xvel = random(-2,2)
ball1tvel = random(-2,2)

#ball2:
ball2x = random(width)
ball2y = random(height)
ball2xvel = random(-2,2)
ball2tvel = random(-2,2)

#ball3:
ball3x = random(width)
ball3y = random(height)
ball3xvel = random(-2,2)
ball3tvel = random(-2,2)

#update ball1:
ball1x += ball1xvel
ball1y += ball1yvel
ellipse(ball1x,ball1y,20,20)

#update ball2:
ball2x += ball2xvel
ball2y += ball2yvel
ellipse(ball2x,ball2y,20,20)

#update ball3:
ball3x += ball3xvel
ball3y += ball3yvel
ellipse(ball3x,ball3y,20,20)
```

Listing 9-6: Creating multiple balls without classes. Way too much code!

This is the code for creating only three balls. As you can see, it's very long, and this doesn't even include the bouncing part! Let's see how we can use classes to make this task easier.

CREATING OBJECTS USING CLASSES

In programming, a class works like a recipe that details a way to create an object with its own specific properties. Using classes, we tell Python how to make a ball once. Then all we have to do is create a bunch of balls using a `for` loop and put them in a list. Lists are great for saving numerous things—strings, numbers, and objects!

Follow these three steps when using classes to create objects:

1. *Write the class.* This is like a recipe for how to make balls, planets, rockets, and so on.
2. *Instantiate the object or objects.* You do this by calling the objects in the `setup()` function.
3. *Update the object or objects.* Do this in the `draw()` function (the display loop).

Let's use these steps to put the code we've already written into a class.

Writing the Class

The first step in creating objects using classes is to write a class that tells the program how to make a ball. Let's add the code in Listing 9-7 at the very beginning of our existing program.

BouncingBall.pyde

```
ballList=[] #empty list to put the balls in

class Ball:
    def __init__(self,x,y):
        '''How to initialize a Ball'''
        self.xcor = x
        self.ycor = y
        self.xvel = random(-2,2)
        self.yvel = random(-2,2)
```

Listing 9-7: Defining a class called `Ball`

Note that because we're putting the position and velocity variables into the `Ball` class as properties, you can delete the following lines from your existing code:

```
xcor = 300
ycor = 300
xvel = 1
yvel = 2
```

In Listing 9-7, we create an empty list we'll use to save the balls in; then we start defining the recipe. The name of a class object, which is `Ball` in this case, is always capitalized. The `__init__` method is a requirement to create a class in Python that contains all the properties the object gets when it's initialized. Otherwise, the class won't work.

The `self` syntax simply means every object has its own methods and properties, which are functions and variables that can't be used except by a `Ball` object. This means that each Ball has its own xcor, its own ycor, and so on. Because we might have to create a Ball at a specific location at some point, we made x and y parameters of the __init__ method. Adding these parameters allows us to tell Python the location of a Ball when we create it, like this:

```
Ball(100,200)
```

In this case, the ball will be located at the coordinate (100,200).

The last lines in Listing 9-7 tell Processing to assign a random number between –2 and 2 to be the x- and y-velocity of the new ball.

Instantiating the Object

Now that we've created a class called Ball, we need to tell Processing how to update the ball every time the draw() function loops. We'll call that the update method and nest it inside the Ball class, just like we did with __init__. You can simply cut and paste all the ball code into the update() method and then add self. to each of the object's properties, as shown in Listing 9-8.

BouncingBall.pyde

```
ballList=[] #empty list to put the balls in

class Ball:
    def __init__(self,x,y):
        '''How to initialize a Ball'''
        self.xcor = x
        self.ycor = y
        self.xvel = random(-2,2)
        self.yvel = random(-2,2)

    def update(self):
        self.xcor += self.xvel
        self.ycor += self.yvel
        #if the ball reaches a wall, switch direction
        if self.xcor > width or self.xcor < 0:
            self.xvel = -self.xvel
        if self.ycor > height or self.ycor < 0:
            self.yvel = -self.yvel
        ellipse(self.xcor,self.ycor,20,20)
```

Listing 9-8: Creating the update() method

Here, we placed all the code for moving and bouncing a ball into the update() method of the Ball class. The only new code is self in the velocity variables, making them velocity properties of the Ball object. Although it might seem like there are many instances of self, that's how we tell Python that the x-coordinate, for example, belongs to that specific ball and not another. Very soon, Python is going to be updating a hundred balls, so we need self to keep track of each one's location and velocity.

Now that the program knows how to create and update a ball, let's update the setup() function to create three balls and put them into the ball list (ballList), as shown in Listing 9-9.

```
def setup():
    size(600,600)
    for i in range(3):
        ballList.append(Ball(random(width),
                             random(height)))
```

Listing 9-9: Creating three balls in the setup() function

We created `ballList` in Listing 9-7 already, and here we're appending to the list a `Ball` at a random location. When the program creates (instantiates) a new ball, it will now choose a random number between 0 and the width of the screen to be the x-coordinate and another random number between 0 and the height of the screen to be the y-coordinate. Then it'll put that new ball into the list. Because we used the loop `for i in range(3)`, the program will add three balls to the ball list.

Updating the Object

Now let's tell the program to go through `ballList` and update all the balls in the list (which means drawing them) every loop using the following `draw()` function:

BouncingBall.pyde

```
def draw():
    background(0) #black
    for ball in ballList:
        ball.update()
```

Note that we still want the background to be black, and then we loop over the ball list and for every ball in the list we run its `update()` method. All the previous code in `draw()` went into the `Ball` class!

When you run this sketch, you should see three balls moving around the screen and bouncing off the walls! The great thing about using classes is that it's super easy to change the number of balls. All you have to do is change the *number* in `for i in range(number):` in the `setup()` function to create even more bouncing balls. When you change this to 20, for example, you'll see something like Figure 9-4.

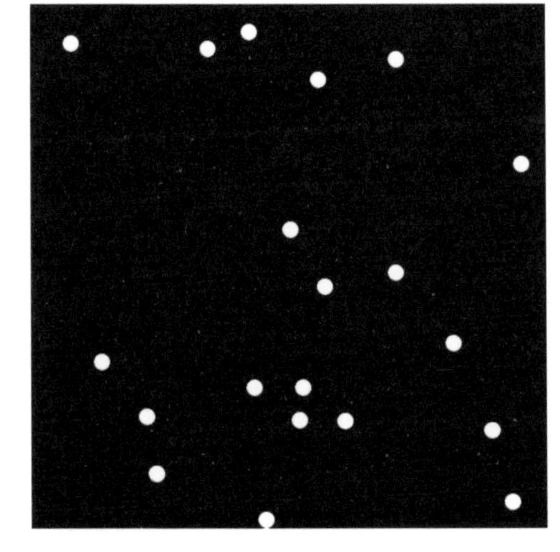

Figure 9-4: Creating as many bouncing balls as you want!

What's cool about using classes is that you can give an object any properties or methods you want. For example, we don't have to make our balls all the same color. Add the three lines of code shown in Listing 9-10 to your existing `Ball` class.

BouncingBall.pyde

```
class Ball:
    def __init__(self,x,y):
        '''How to initialize a Ball'''
        self.xcor = x
        self.ycor = y
        self.xvel = random(-2,2)
        self.yvel = random(-2,2)
        self.col = color(random(255),
                         random(255),
                         random(255))
```

Listing 9-10: Updating the `Ball` *class*

This code gives every ball its own color when it's created. Processing's `color()` function needs three numbers that represent red, green, and blue, respectively. RGB values go from 0 to 255. Using `random(255)` lets the program choose the numbers randomly, resulting in a randomly chosen color. However, because the `__init__` method runs only one time, once the ball has a color, it keeps it.

Next, in the `update()` method, add the following line so the ellipse gets filled with its own randomly chosen color:

```
        fill(self.col)
        ellipse(self.xcor,self.ycor,20,20)
```

Before a shape or line gets drawn, you can declare its color using `fill` for shapes or `stroke` for lines. Here, we tell Processing to use the ball's own color (using `self`) to fill in the following shape.

Now when you run the program, each ball should have a random color, as shown in Figure 9-5!

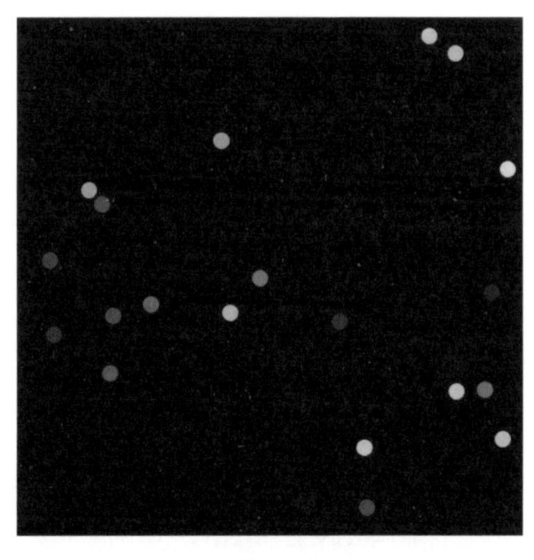

Figure 9-5: Giving balls their own colors

> **EXERCISE 9-1:**
> **CREATING BALLS OF DIFFERENT SIZES**
>
> Give each ball its own size, between 5 and 50 units.

GRAZING SHEEP PROGRAM

Now that you can create classes, let's make something useful. We'll code a Processing sketch of an ecosystem that simulates sheep walking around eating grass. In this sketch, the sheep have a certain level of energy that gets depleted as they walk around, and their energy gets replenished when they eat grass. If they get enough energy, they reproduce. If they don't get enough energy, they die. We could potentially learn a lot about biology, ecology, and evolution by creating and tweaking this model.

In this program, the `Sheep` objects are kind of like the `Ball` objects you created earlier in this chapter; each has its own x- and y-position and size, and is represented by a circle.

WRITING THE CLASS FOR THE SHEEP

Start a new Processing sketch and save it as *SheepAndGrass.pyde*. First, let's create a class that makes a `Sheep` object with its own x- and y-position and its own size. Then we'll create an `update` method that draws an ellipse representing the sheep's size at the sheep's location.

The class code is nearly identical to the `Ball` class, as you can see in Listing 9-11.

SheepAndGrass.pyde

```
class Sheep:
    def __init__(self,x,y):
        self.x = x #x-position
        self.y = y #y-position
        self.sz = 10 #size

    def update(self):
        ellipse(self.x,self.y,self.sz,self.sz)
```

Listing 9-11: Creating a class for one sheep

Because we know we'll be making a bunch of sheep, we start off creating a `Sheep` class. In the required `__init__` method, we set the x- and y-coordinates of the sheep to the parameters we'll declare when creating a sheep instance. I've set the size of the sheep (the diameter of the ellipse) to 10 pixels, but you can have bigger or smaller sheep if you like. The `update()` method simply draws an ellipse of the sheep's size at the sheep's location.

Here's the `setup()` and `draw()` code for a Processing sketch containing one `Sheep`, which I've named `shawn`. Add the code shown in Listing 9-12 right below the `update()` method you just wrote in Listing 9-11.

```
def setup():
    global shawn
    size(600,600)
    #create a Sheep object called shawn at (300,200)
    shawn = Sheep(300,200)

def draw():
    background(255)
    shawn.update()
```

Listing 9-12: Creating a `Sheep` object named shawn

We first create `shawn`, an instance of a `Sheep` object, in the `setup()` function. Then we update it in the `draw()` function—but Python doesn't know we mean the same `shawn` unless we tell it that `shawn` is a global variable.

When you run this code, you should see something like what's shown in Figure 9-6.

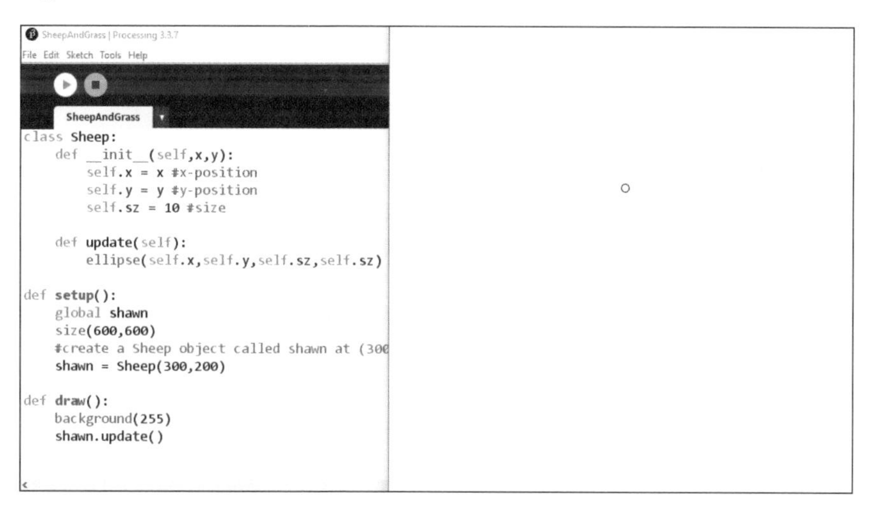

Figure 9-6: One sheep

You get a white screen with a little circular sheep at the coordinate (300,200), which is 300 pixels to the right of the starting point and 200 pixels down.

PROGRAMMING SHEEP TO MOVE AROUND

Now let's teach a `Sheep` how to move around. We'll start by programming the `Sheep` to move around randomly. (You can always program it to move differently in the future if you want to.) Listing 9-13 changes the x- and y-coordinates of a `Sheep` by a random number between –10 and 10. Return to your existing code and add the following lines above the ellipse() function within the update() method:

SheepAndGrass.pyde

```
    def update(self):
        #make sheep walk randomly
        move = 10 #the maximum it can move in any direction
        self.x += random(-move, move)
        self.y += random(-move, move)
        fill(255) #white
        ellipse(self.x,self.y,self.sz,self.sz)
```

Listing 9-13: Making the sheep move randomly

This code creates a variable called move to specify the maximum value or distance the sheep will be able to move on the screen. Then we set move to 10 and use it to update the sheep's x- and y-coordinates by a random number between -move (–10) and move (10). Finally, we use fill(255) to set the sheep's color to white for now.

When you run this code, you should see the sheep wandering around randomly—and it might wander off the screen.

Let's give the sheep some company. If we want to create and update more than one object, it's a good idea to put them in a list. Then in the draw() function, we'll go through the list and update each Sheep. Update your existing code to look like Listing 9-14.

SheepAndGrass.pyde

```
class Sheep:
    def __init__(self,x,y):
        self.x = x #x-position
        self.y = y #y-position
        self.sz = 10 #size

    def update(self):
        #make sheep walk randomly
        move = 10 #the maximum it can move in any direction
        self.x += random(-move, move)
        self.y += random(-move, move)
        fill(255) #white
        ellipse(self.x,self.y,self.sz,self.sz)

sheepList = [] #list to store sheep

def setup():
    size(600,600)
    for i in range(3):
        sheepList.append(Sheep(random(width),
                               random(height)))
def draw():
    background(255)
    for sheep in sheepList:
        sheep.update()
```

Listing 9-14: Creating more sheep using a for loop

This code is similar to the code we wrote to put the bouncing balls in a list. First, we create a list to store the sheep. Then we create a for loop and put a Sheep in the sheep list. Then in the draw() function, we write another for loop to go through the sheep list and update each one according to the update() method we already defined. When you run this code, you should get three Sheep walking around randomly. Change the number 3 in for i in range(3): to a larger number to add even more sheep.

CREATING THE ENERGY PROPERTY

Walking takes up energy! Let's give the sheep a certain level of energy when they're created and take away their energy when they walk. Use the code in Listing 9-15 to update your existing __init__ and update() methods in the *SheepAndGrass.pyde*.

```
class Sheep:
    def __init__(self,x,y):
        self.x = x #x-position
        self.y = y #y-position
        self.sz = 10 #size
        self.energy = 20 #energy level

    def update(self):
        #make sheep walk randomly
        move = 1
        self.energy -= 1 #walking costs energy
        if sheep.energy <= 0:
            sheepList.remove(self)
        self.x += random(-move, move)
        self.y += random(-move, move)
        fill(255) #white
        ellipse(self.x,self.y,self.sz,self.sz)
```

Listing 9-15: Updating __init__ and update() with the energy property

We do this by creating an energy property in the __init__ method and set it to 20, the energy level every sheep starts with. Then self.energy -= 1 in the update() method lowers the sheep's energy level by 1 when it walks around.

Then we check whether the sheep is out of energy, and if it is, we remove it from the sheepList. Here, we use a conditional statement to check whether if sheep.energy <= 0 returns True. If so, we remove that sheep from the sheepList using the remove() function. Once that Sheep instance is gone from the list, it doesn't exist anymore.

CREATING GRASS USING CLASSES

When you run the program, you should see the Sheep move around for a second and then disappear—walking around is costing the sheep energy, and once that energy is gone, the sheep dies. What we need to do is to give the sheep grass to eat. We'll call each patch of grass Grass and make a new class for it. Grass will have its own x- and y-value, size, and energy content. We'll also make it change color when it's eaten.

In fact, we'll be using a bunch of different colors in this sketch for our sheep and our grass, so let's add the code in Listing 9-16 to the very beginning of the program so we can just refer to the colors by their names. Feel free to add other colors too.

```
WHITE = color(255)
BROWN = color(102,51,0)
RED = color(255,0,0)
GREEN = color(0,102,0)
YELLOW = color(255,255,0)
PURPLE = color(102,0,204)
```

Listing 9-16: Setting colors as constants

Using all-caps for the color names indicates that they're constants and won't change in value, but that's just for the programmer. There's nothing inherently magical about the constants, and you can change these values if you want. Setting constants lets you just type the names of the colors instead of having to write the RGB values every time. We'll do this when we make the grass green. Update your existing code by adding the code in Listing 9-17 right after the Sheep class in *SheepAndGrass.pyde*:

```
class Grass:
    def __init__(self,x,y,sz):
        self.x = x
        self.y = y
        self.energy = 5 #energy from eating this patch
        self.eaten = False #hasn't been eaten yet
        self.sz = sz

    def update(self):
        fill(GREEN)
        rect(self.x,self.y,self.sz,self.sz)
```

Listing 9-17: Writing the Grass *class*

You're probably starting to get used to the structure of the class notation. It conventionally starts with the __init__ method, where you create its properties. In this case, you tell the program that Grass will have an x- and y-location, an energy level, a Boolean (True/False) variable that keeps track of whether the grass has been eaten or not, and a size. To update a patch of grass, we just create a green rectangle at the Grass object's location.

Now we have to initialize and update our grass, the same way we did for our sheep. Because there will be a lot of grass, let's create a list for it. Before the setup() function, add the following code.

```
sheepList = [] #list to store sheep
grassList = [] #list to store grass
patchSize = 10 #size of each patch of grass
```

We might want to vary the size of the patch of grass in the future, so let's create a variable called `patchSize` so we'll only have to change it in one place. In the `setup()` function, after creating the sheep, create the grass by adding the new code in Listing 9-18.

```
def setup():
    global patchSize
    size(600,600)
    #create the sheep
    for i in range(3):
        sheepList.append(Sheep(random(width),
                               random(height)))
    #create the grass:
    for x in range(0,width,patchSize):
        for y in range(0,height,patchSize):
            grassList.append(Grass(x,y,patchSize))
```

Listing 9-18: Updating the `Grass` object using `patchSize` variable

In this example, `global patchSize` tells Python that we're using the same `patchSize` variable everywhere. Then we write two `for` loops (one for x and the other for y) to append `Grass` to the grass list so we can create a square grid of grass.

Then we update everything in the `draw()` function, just like we did for the sheep. Whatever is drawn first will be drawn covered up by what's drawn after, so we'll update the grass first by changing the `draw()` function to the code in Listing 9-19.

SheepAndGrass.pyde

```
def draw():
    background(255)
    #update the grass first
    for grass in grassList:
        grass.update()
    #then the sheep
    for sheep in sheepList:
        sheep.update()
```

Listing 9-19: Updating the grass before the sheep

When you run this code, you should see a grid of green squares, like in Figure 9-7.

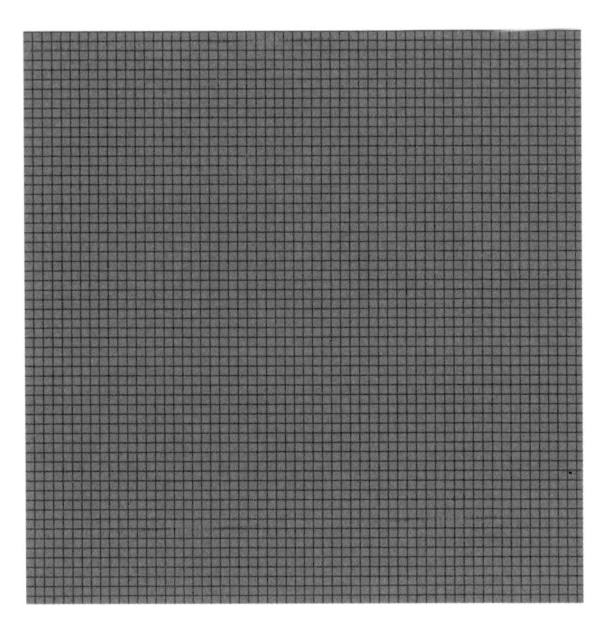

Figure 9-7: Grass with grid lines

Let's shut off the black outline so it'll look like a smooth field of grass. Add `noStroke()` to the `setup()` function to remove the outline of the green squares:

```
def setup():
    global patchSize
    size(600,600)
    noStroke()
```

Now we have our grass!

MAKING THE GRASS BROWN WHEN EATEN

How do we make it so that when a sheep is on a patch of grass, the sheep gets the grass's energy and the patch of grass turns brown to show that the sheep has eaten it? Change the `update()` method for `Grass` by adding the following lines of code:

```
    def update(self):
        if self.eaten:
            fill(BROWN)
        else:
            fill(GREEN)
        rect(self.x,self.y,self.sz,self.sz)
```

This code tells Processing that if the patch of grass is "eaten," the rectangle should be filled with a brown color. Otherwise, the grass should be colored green. There's more than one way for a sheep to "eat" grass. One way is to make each patch of grass check the entire `sheepList` for a sheep on its location, which could mean tens of thousands of patches are checking thousands of sheep. Those numbers could get big. However, because each patch of grass is in the grassList, an alternate way is that when a sheep changes its location, it could simply change the patch at that location to "eaten" (if it isn't already) and get energy from eating it. That would mean a lot less checking.

The problem is that the x- and y-coordinates of the sheep don't exactly match up to where the patches of grass are in the grassList. For example, our patchSize is 10, meaning that if a sheep is at (92,35), it'll be on the 10th patch to the right and the 4th patch down (because the "first" patch is from x = 0 to x = 9). We're dividing by the patchSize to get the "scaled" x- and y-values, 9 and 3.

However, the grassList doesn't have rows and columns. We do know that the x-value, 9, means it's the 10th row (don't forget row 0), so we'll just have to add in nine rows of 60 (the height divided by the patchSize) and then add the y-value to get the index of the patch of grass the sheep is on. Therefore, we need a variable to tell us how many patches of grass there are in a row, which we'll call rows_of_grass. Add global rows_of_grass to the beginning of the setup() function and then add this line to setup() after declaring the size:

```
rows_of_grass = height/patchSize
```

This takes the width of the display window and divides it by the size of the patches of grass to tell us how many columns of grass there are. The code to add to the Sheep class is in Listing 9-20.

SheepAndGrass.pyde

```
    self.x += random(-move, move)
    self.y += random(-move, move)
    #"wrap" the world Asteroids-style
❶   if self.x > width:
        self.x %= width
    if self.y > height:
        self.y %= height
    if self.x < 0:
        self.x += width
    if self.y < 0:
        self.y += height
    #find the patch of grass you're on in the grassList:
❷   xscl = int(self.x / patchSize)
    yscl = int(self.y / patchSize)
❸   grass = grassList[xscl * rows_of_grass + yscl]
    if not grass.eaten:
        self.energy += grass.energy
        grass.eaten = True
```

Listing 9-20: Updating the sheep's energy level and turning the grass brown

After updating the sheep's location, we "wrap" the coordinates ❶ so if the sheep walks off the screen in one direction, it shows up on the other side of the screen, like in the video game *Asteroids*. We calculate which patch the sheep is on according to the patchSize ❷. Then we use code to go from x- and y-values to the index of that patch in the grassList ❸. We now know the exact index of the patch of grass the sheep is on. If this patch of grass is not already eaten, the sheep eats it! It gets the energy from the grass, and the grass's eaten property is set to True.

Run this code, and you'll see the three sheep running around eating grass, which turns brown once it's eaten. Slow the sheep down by changing the move variable to a lesser value, such as 5. You can also scale down the patches by changing one number, the patchSize variable, to 5. Try other values if you like.

Now we can create more Sheep. Let's change the number in the for i in range line to 20, like so:

```
#create the sheep
for i in range(20):
    sheepList.append(Sheep(random(width),
                           random(height)))
```

When you run this code, you should see something like Figure 9-8.

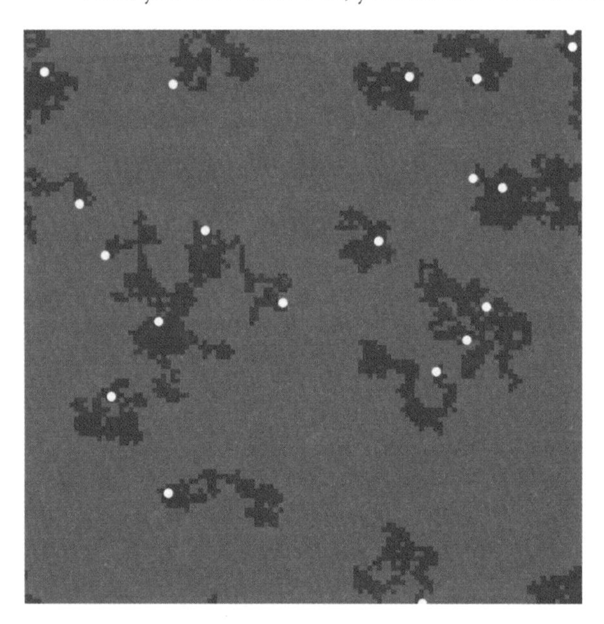

Figure 9-8: A herd of sheep!

Now there are 20 sheep walking around, leaving patches of brown grass.

GIVING EACH SHEEP A RANDOM COLOR

Let's have the sheep choose a color when they're "born." After the code defining the color constants, let's put some colors into a color list, like this:

```
YELLOW = color(255,255,0)
PURPLE = color(102,0,204)
colorList = [WHITE,RED,YELLOW,PURPLE]
```

Make the following changes to the Sheep class to use different colors. First, you need to give Sheep a color property. Because color is already a keyword in Processing, col is used in Listing 9-21.

```
class Sheep:
    def __init__(self,x,y,col):
        self.x = x #x-position
        self.y = y #y-position
        self.sz = 10 #size
        self.energy = 20
        self.col = col
```

Listing 9-21: Adding a color property to the Sheep class

Then in the update() method, replace the fill line with this:

```
        fill(self.col) #its own color
        ellipse(self.x,self.y,self.sz,self.sz)
```

Before the ellipse is drawn, fill(self.col) tells Processing to fill the ellipse with the Sheep's own randomly chosen color.

When all the Sheep are instantiated in the setup() function, you need to give them a random color. That means at the top of the program you have to import the choice() function from the random module, like this:

```
from random import choice
```

Python's choice() function allows you to have one item chosen at random from a list and then returned. We should be able to tell the program to do this as follows:

```
choice(colorList)
```

Now the program will return a single value from the color list. Finally, when you're creating the Sheep, add the random choice of color from the color list as one of the arguments you pass to the Sheep constructor, as shown here:

```
def setup():
    size(600,600)
    noStroke()
    #create the sheep
    for i in range(20):
        sheepList.append(Sheep(random(width),
                               random(height),
                               choice(colorList)))
```

Now when you run this code, you should see a bunch of randomly colored sheep walking around the screen, as shown in Figure 9-9.

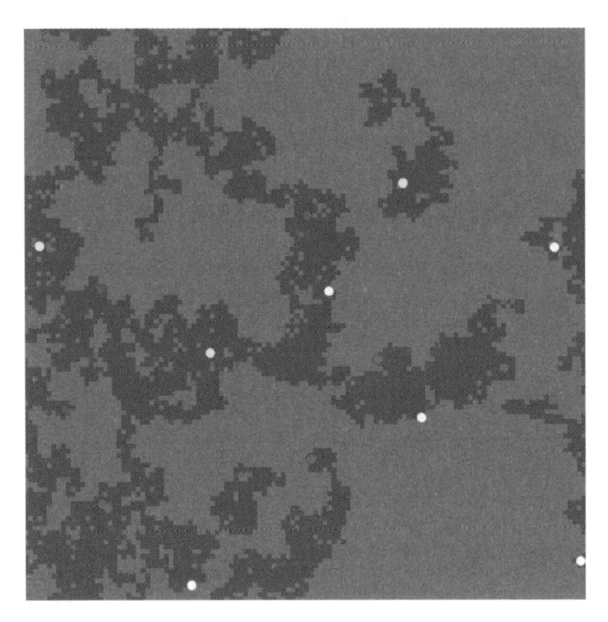

Figure 9-9: Multicolored sheep

Each new sheep gets assigned one of the four colors we defined in `colorList`: white, red, yellow, or purple.

PROGRAMMING SHEEP TO REPRODUCE

Unfortunately, in our current program the sheep eat the grass until they wander too far away from the grass, run out of energy, and die. To prevent this, let's tell the sheep to use some of that energy to reproduce.

Let's use the code in Listing 9-22 to tell the sheep to reproduce if their energy level reaches 50.

```
if self.energy <= 0:
    sheepList.remove(self)
if self.energy >= 50:
    self.energy -= 30 #giving birth takes energy
    #add another sheep to the list
    sheepList.append(Sheep(self.x,self.y,self.col))
```

Listing 9-22: Adding a conditional for sheep to reproduce

The conditional `if self.energy >= 50:` checks whether that sheep's energy is greater than or equal to 50. If it is, we decrement the energy level by 30 for birthing and add another sheep to the sheep list. Notice that the new sheep is at the same location and is the same color as its parent. Run this code, and you should see the sheep reproduce, like in Figure 9-10.

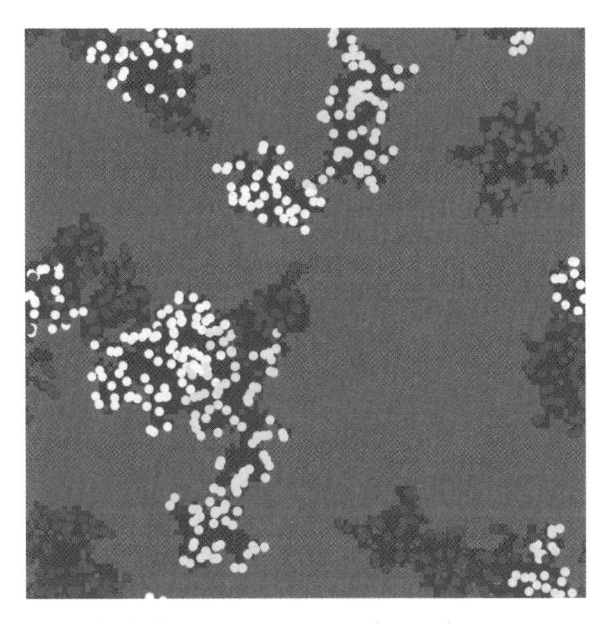

Figure 9-10: Sheep eating grass and reproducing

Soon you should see what looks like tribes of similarly colored sheep.

LETTING THE GRASS REGROW

Unfortunately, the sheep soon eat up all the grass in their area and die (probably a lesson in there somewhere). We need to allow our grass to regrow. To do this, change the Grass's `update()` method to this:

```
def update(self):
    if self.eaten:
        if random(100) < 5:
            self.eaten = False
        else:
            fill(BROWN)
    else:
        fill(GREEN)
    rect(self.x,self.y,self.sz,self.sz)
```

The Processing code `random(100)` generates a random number between 0 and 100. If the number is less than 5, we regrow a patch of grass by setting its `eaten` property to `False`. We use the number 5 because this gives us a probability of 5/100 that eaten grass will regrow during each frame. Otherwise, it stays brown.

Run the code, and you should see something like Figure 9-11.

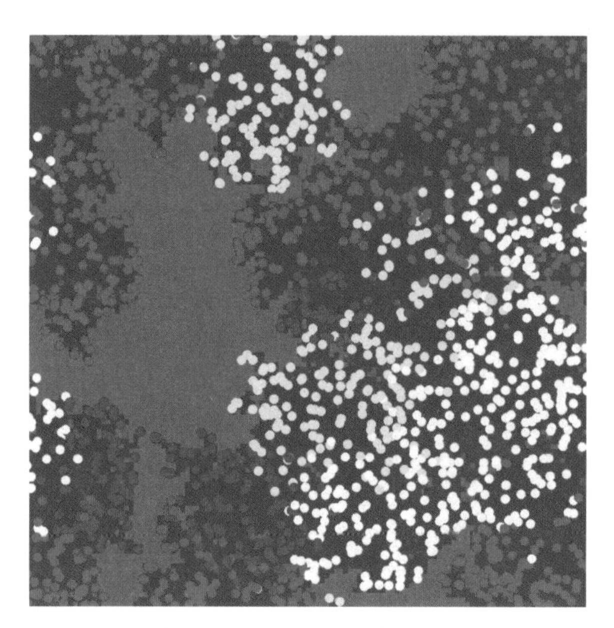

Figure 9-11: The grass regrows and the sheep populate the whole screen!

Now you might get so many sheep that the program starts to slow down! This could be because the sheep have too much energy. If so, try reducing the amount of energy each patch of grass contains from 5 to 2:

```
class Grass:
    def __init__(self,x,y,sz):
        self.x = x
        self.y = y
        self.energy = 2 #energy from eating this patch
        self.eaten = False #hasn't been eaten yet
        self.sz = sz
```

That seems to be a good balance that lets the sheep population grow at a reasonable pace. Play around with the numbers all you want—it's your world!

PROVIDING AN EVOLUTIONARY ADVANTAGE

Let's give one of the sheep groups an advantage. You can choose any advantage you can think of (getting more energy from grass or producing more offspring at a time, for instance). For this example, we're going to let the purple sheep walk a little further than the others. Will that make any difference? To find out, make the `Sheep`'s `update()` method match the following code:

```
    def update(self):
        #make sheep walk randomly
```

```
move = 5 #the maximum it can move in any direction
if self.col == PURPLE:
    move = 7
self.energy -= 1
```

This conditional checks whether the Sheep's color is purple. If so, it sets the Sheep's move value to 7. Otherwise, it leaves the value at 5. This allows the purple sheep to travel further, and therefore more likely to find green patches, than the other sheep. Let's run the code and check the outcome, which should look like Figure 9-12.

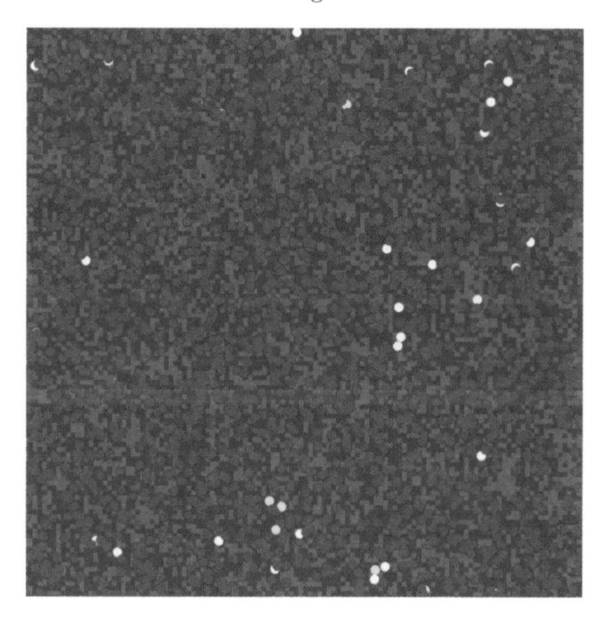

Figure 9-12: Giving purple sheep an advantage

After a little while it sure looks like that tiny advantage paid off for the purple sheep. They're dominating the environment and pushing out all the other sheep just by competing for grass. This simulation could spark interesting discussions about ecology, invasive species, biodiversity, and evolution.

> **EXERCISE 9-2: SETTING SHEEP LIFESPAN**
>
> Create an "age" property and decrease it every time the sheep update so they live for only a limited amount of time.

EXERCISE 9-3: CHANGING SHEEP SIZE

Vary the size of the sheep according to their energy level.

SUMMARY

In this chapter, you learned how to make objects using classes, which involved defining the class using properties and then instantiating ("creating") and updating the object. This let you create multiple similar-but-independent objects with the same properties more efficiently. The more you use classes, the more creative you can get by making autonomous objects walk, fly, or bounce around without your having to code every step!

Knowing how to use classes supercharges your coding abilities. Now you can create models of complicated situations easily, and once you tell the program what to do with one particle, or planet, or sheep, it'll be able to make a dozen, a hundred, or even a million of them very easily!

You also got a taste of setting up models to explore physical, biological, chemical, or environmental situations with very few equations! A physicist once told me that's often the most efficient method for solving problems involving many factors, or "agents." You set up a computer model, let it run, and look at the results.

In the next chapter, you'll learn how to create fractals using an almost-magical phenomenon called recursion.

10

CREATING FRACTALS USING RECURSION

What's another word for thesaurus*?*
—Steven Wright

Fractals are delightfully complicated designs, where each smaller part of the design contains the entire design (see Figure 10-1). They were invented (or discovered, since fractals exist in nature) by Benoit Mandelbrot in 1980 when he was visualizing some complex functions on a state-of-the-art IBM computer.

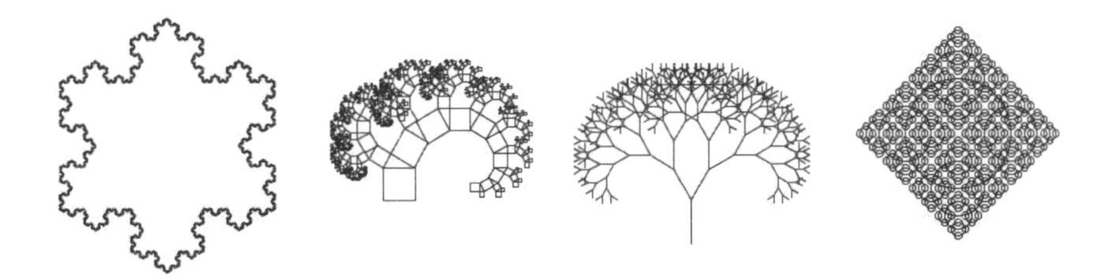

Figure 10-1: Examples of fractals

Fractals don't look like regular shapes we recognize from geometry, like squares, triangles, and circles. Their shapes are crooked and jagged, making them great models for simulating natural phenomena. In fact, scientists use fractals to model everything from the arteries in your heart, to earthquakes, to neurons in your brain.

What makes fractals so interesting is that they illustrate how you can get surprisingly complex designs from simple rules being run over and over and patterns being repeated at smaller and smaller scale.

Our main interest is the interesting, complicated designs you can make using fractals. There's a picture of a fractal in every math book these days, but textbooks never show you how to make one—you need a computer to do that. In this chapter, you learn how to make your own fractals using Python.

THE LENGTH OF A COASTLINE

Before you can start creating fractals, let's look at a simple example to understand how fractals can be useful. A mathematician named Lewis Richardson asked a simple question; "How long is the coastline of England?" As you can see in Figure 10-2, the answer depends on how long your ruler is.

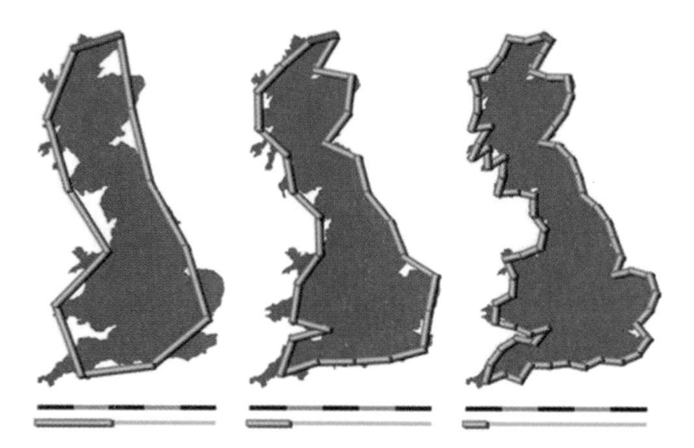

Figure 10-2: Approximating the length of a coastline

The smaller your ruler, the more closely you can approximate the coastline's jagged edges, which means you'll end up with a longer measurement. The cool thing is that *the length of the coastline approaches infinity as the length of the ruler gets close to zero!* This is known as the Coastline Paradox.

Think this is just abstract mathematical noodling? Coastline length estimates can vary wildly in the real world. Even with modern technology, it all depends on the scale used to measure the map. We'll draw a figure like Figure 10-3, the Koch snowflake, to show how a fractal can prove a rough enough coastline can get as long as you want!

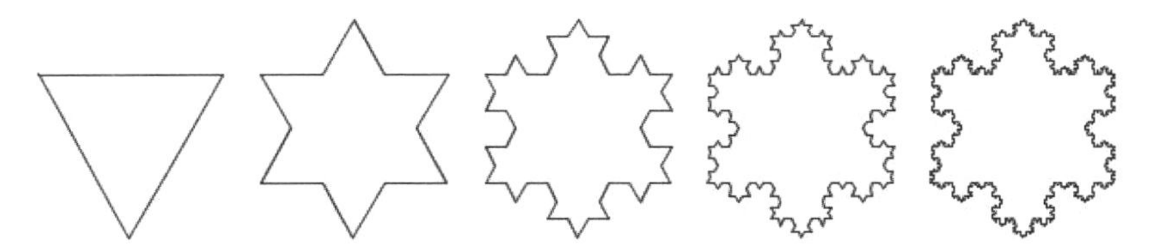

Figure 10-3: An increasingly detailed fractal, modeling an increasingly rough coastline

First, you're going to need to learn a few tricks, like recursion.

WHAT IS RECURSION?

The power of fractals is that you can repeat patterns of numbers or shapes that get smaller at every step until you're dealing with very small numbers. The key to repeating all this code is a concept called *recursion*, which is when something is defined in terms of itself. Some of these jokes illustrate how recursion works:

- If you google "recursion," it asks you, "Did you mean *recursion*?"
- In the index to more than one computer programming book, there's an entry like this: "recursion, see *recursion*."

As you can imagine, recursion is a pretty strange concept. The virtue of recursion is that it can tidy up code that would otherwise be too complicated, but the disadvantage is that you can end up using up too much memory.

WRITING THE FACTORIAL() FUNCTION

Let's see recursion in action by writing a function for the factorial of a number. You may recall from math class that the *factorial* of n (expressed as $n!$) is defined as the product of all the integers from 1 to n. For example, $5! = 1 \times 2 \times 3 \times 4 \times 5 = 120$.

The formula looks like this: $n! = 1 \times 2 \times 3 \ldots \times (n-2) \times (n-1) \times n$. This is an example of a recursive sequence, because $5! = 5 \times 4!$ and $4! = 4 \times 3!$, and so on. Recursion is an important concept in math because math is all about patterns, and recursion allows you to copy and extend patterns infinitely!

We can define the factorial of n as the product of n and the factorial of n – 1. We just have to define the factorial of 0 (which is 1, not 0) and the factorial of 1 and then use a recursive statement. Open a new file in IDLE, save it as *factorial.py*, and then enter with the code in Listing 10-1.

factorial.py

```
def factorial(n):
    if n == 0:
        return 1
    else:
        return n * factorial(n - 1)
```

Listing 10-1: Using a recursive statement to write the `factorial()` function

First, we're saying, "If the user (or the program) asks for the factorial of 0 or 1, return 1." This is because 0! and 1! both equal 1. Then we tell the program, "For any other number *n*, return *n* times the factorial of the number 1 less than *n*."

Notice that on the last line of Listing 10-1, we're calling the `factorial()` function *inside* the definition of the `factorial()` function! That's like a recipe for a loaf of bread containing the step "Bake a loaf of bread." People wouldn't even begin following a recipe written like that. But computers can start going through the steps and follow them throughout the process.

In this example, when we ask for the factorial of 5, the program proceeds obediently and makes it to the last line, where it asks for the factorial of $n - 1$, which in this case (because $n = 5$) is the factorial of 4. To calculate factorial $(5 - 1)$, the program starts the `factorial()` function again with $n = 4$ and tries to evaluate the factorial of 4 the same way, followed by the factorial of 3, the factorial of 2, the factorial of 1, and finally the factorial of 0. Because we already defined the function to return the factorial of 0 as 1, the function can go back up through the process, evaluating the factorial of 1, then 2, then 3, then 4, and finally 5.

Defining a function recursively (by calling the function inside its own definition) might seem confusing, but it's the key to making all the fractals in this chapter. Let's start with a classic: the fractal tree.

BUILDING A FRACTAL TREE

Making a fractal starts with defining a simple function and adding a call to the function inside the function itself. Let's try building a fractal tree that looks like Figure 10-4.

Figure 10-4: A fractal tree

This would be an incredibly complicated design to create if you had to tell the program every line to draw. But it takes surprisingly little code if you use recursion. Using translations, rotations, and the `line()` function, we'll first draw a Y in Processing, as shown in Figure 10-5.

Figure 10-5: The beginnings of a fractal tree

The only requirement to eventually make this Y into a fractal is that after the program draws the Y tree, along with the branches, the program has to return to the bottom of the "trunk." This is because the "branches" are going to become Y's themselves. If the program doesn't return to the bottom of the Y every time, we won't get our tree.

Writing the y() Function

Your Y doesn't have to be perfect or symmetrical, but here's my code for drawing a Y. Open a new sketch in Processing, name it *fractals.pyde*, and enter the code in Listing 10-2.

fractals.pyde

```
def setup():
    size(600,600)

def draw():
    background(255)
    translate(300,500)
    y(100)

def y(sz):
    line(0,0,0,-sz)
    translate(0,-sz)
    rotate(radians(30))
    line(0,0,0,-0.8*sz) #right branch
    rotate(radians(-60))
    line(0,0,0,-0.8*sz) #left branch
    rotate(radians(30))
    translate(0,sz)
```

Listing 10-2: Writing the `y()` *function for the fractal tree*

We set up the Processing sketch the way we always do: in the `setup()` function we tell the program what size to make the display window, and then in the `draw()` function we set the background color (255 is white) and translate to where we want to start drawing. Finally, we call the `y()` function and pass the number 100 for the size of the "trunk" of the fractal tree.

The `y()` function takes a number `sz` as a parameter to be the length of the trunk of the tree. Then all the branches will be based on that number. The first line of code in the `y()` function draws the trunk of the tree using a vertical line. To create a line branching off to the right, we translate the vertical line up the trunk of the tree (in the negative y-direction) and then rotate it 30 degrees to the right. Next, we draw another line for the right branch, rotate to the left (negative 60 degrees), and draw another line for the left branch. Finally, we have to rotate so we're facing straight up again so that we can translate down the trunk again. Save and run this sketch, and you should see the Y in Figure 10-5.

We can convert this program that draws a single Y into one that draws a fractal by making the branches into *smaller* Y's. But if we simply replace "line" with "y" in the `y()` function, our program will get stuck in an infinite loop, throwing an error like this:

```
RuntimeError: maximum recursion depth exceeded
```

Recall that we didn't call `factorial(n)` inside the factorial function but rather called `factorial(n-1)`. We have to introduce a `level` parameter to the `y()` function. Then each branch up, the tree will be a level down, so the branch will get the parameter `level - 1`. This means the trunk is always the highest numbered level and the last set of branches up the tree is always level 0. Here's how to change the `y()` function in Listing 10-3.

fractals.pyde

```
def setup():
    size(600,600)

def draw():
    background(255)
    translate(300,500)
    y(100,2)

def y(sz,level):
    if level > 0:
        line(0,0,0,-sz)
        translate(0,-sz)
        rotate(radians(30))
        y(0.8*sz,level-1)
        rotate(radians(-60))
        y(0.8*sz,level-1)
        rotate(radians(30))
        translate(0,sz)
```

Listing 10-3: Adding recursion to the y() function

Notice that we replaced all the `line()` functions in the code with `y()` functions to draw the branches. Because we changed the call to the `y()` function in `draw()` to `y(100,2)`, we'll get a tree of trunk size 100 with two levels. Try a three-level tree, a four-level one, and so on! You should see something like Figure 10-6.

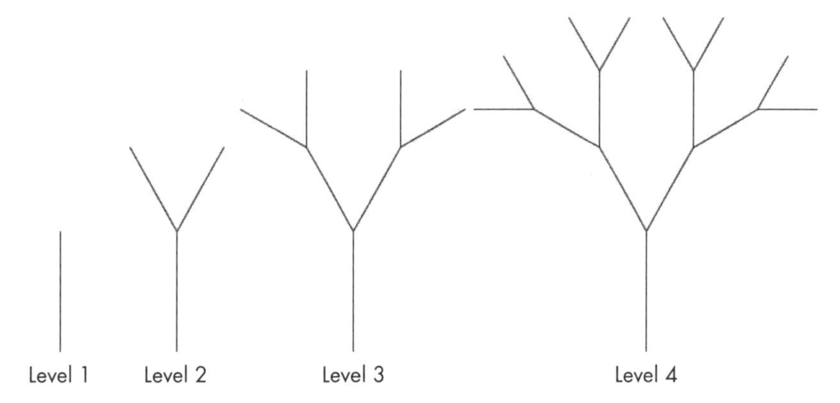

Figure 10-6: Trees of levels 1 through 4

Mapping the Mouse

Now let's make a program that allows you to control the shape of the fractal in real time, just by moving your mouse up or down! We can vary the level of rotation dynamically by tracking the mouse and returning a value between 0 and 10 based on its location. Update the `draw()` function with the code in Listing 10-4.

fractals.pyde

```
def draw():
    background(255)
    translate(300,500)
    level = int(map(mouseX,0,width,0,10))
    y(100,level)
```

Listing 10-4: Adding the `level` parameter to the `draw()` function

Our mouse's x-value can be anywhere between 0 and the width of the window. The `map()` function replaces one range of values with another. In Listing 10-4, `map()` will take the x-value and instead of the output being between 0 and 600 (the width of the display screen), it will be between 0 and 10, the range of levels we want to draw. So we assign that value to a variable called `level` and pass that value to the `y()` function in the next line.

Now that we've tweaked the `draw()` function to return a value based on the position of the mouse, we can vary the shape of our tree by linking the y-coordinate of the mouse to the angle we're rotating by.

The angle of rotation should only go up to 180 because the tree will "fold up" completely at 180 degrees, but the mouse's y-value can go up to 600 since that's the height of the screen we declared in `setup()`. We could do a little math to convert the values ourselves, but it would be easier to just use Processing's built-in `map()` function. We tell the `map()` function what variable we want to map, specifying its current minimum and maximum values and the desired minimum and maximum values. The entire code for the Y fractal tree is shown in Listing 10-5.

fractals.pyde

```
def setup():
    size(600,600)

def draw():
    background(255)
    translate(300,500)
    level = int(map(mouseX,0,width,0,15))
    y(100,level)

def y(sz,level):
    if level > 0:
        line(0,0,0,-sz)
        translate(0,-sz)
        angle = map(mouseY,0,height,0,180)
        rotate(radians(angle))
        y(0.8*sz,level-1)
        rotate(radians(-2*angle))
        y(0.8*sz,level-1)
        rotate(radians(angle))
        translate(0,sz)
```

Listing 10-5: The entire code to make a dynamic fractal tree

We take the mouse's y-value and convert it to a range between 0 and 180 (if you already think in radians, you can map it to between 0 and pi). In the `rotate()` lines, we give it that angle (which is in degrees) and have Processing convert the degrees to radians. The first `rotate()` line will rotate to the right. The second `rotate()` line will rotate a negative angle, meaning to the left. It'll rotate twice as much to the left. Then the third `rotate()` line will rotate to the right again.

When you run the code, you should see something like Figure 10-7.

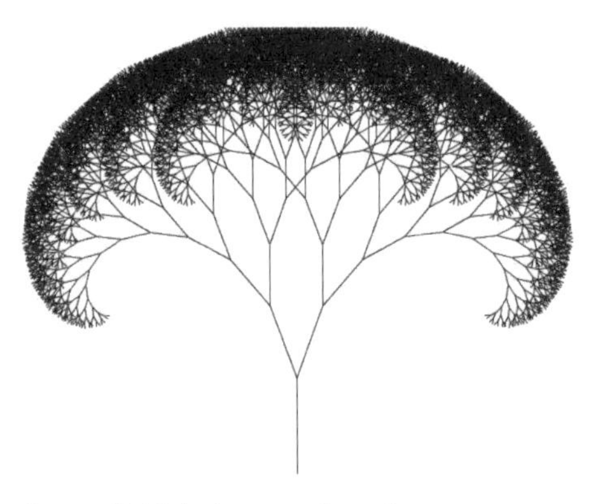

Figure 10-7: A dynamic fractal tree

Now when you move the mouse up or down, left or right, the level and shape of the fractal should change accordingly.

Through drawing the fractal tree, you learned how to use recursion to draw complicated designs using a surprisingly small amount of code. Now we'll return to the coastline problem. How could a coastline, or any line, double or triple in length just from getting more jagged?

KOCH SNOWFLAKE

The Koch snowflake is a famous fractal named after Swedish mathematician Helge von Koch, who wrote about the shape in a paper in 1904! It's made from an equilateral triangle. We start with a line and add a "bump" to it. Then, we add a smaller bump to each resulting line segment and repeat the process, like in Figure 10-8.

Figure 10-8: Adding a "bump" to each segment

Let's start a new Processing sketch, call it *snowflake.pyde*, and add the code in Listing 10-6, which will give us an upside-down equilateral triangle.

snowflake.pyde

```
def setup():
    size(600,600)

def draw():
    background(255)
    translate(100,100)
    snowflake(400,1)

def snowflake(sz,level):
    for i in range(3):
        line(0,0,sz,0)
        translate(sz,0)
        rotate(radians(120))
```

Listing 10-6: Writing the snowflake() function

In the `draw()` function, we call the `snowflake()` function, which for now takes only two parameters: `sz` (the size of the initial triangle) and `level` (the level of the fractal). The `snowflake()` function draws a triangle by starting a loop that repeats the code three times. Inside the loop we draw a line of length `sz`, which will be the side of the triangle, and then translate along the line to the next vertex of the triangle and rotate 120 degrees. Then we draw the next side of the triangle.

When you run the code in Listing 10-6, you should see Figure 10-9.

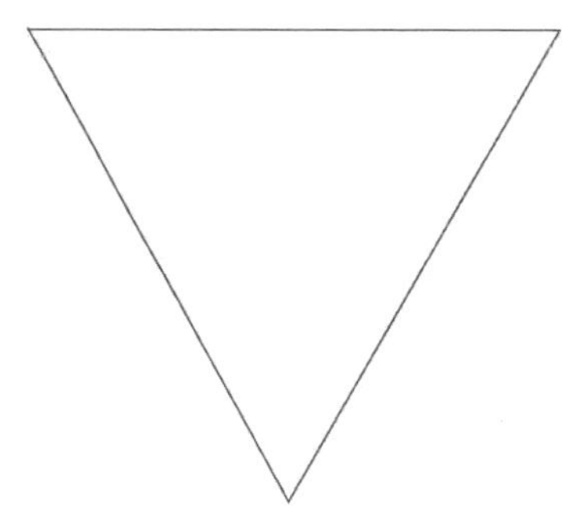

Figure 10-9: Level 1 snowflake: a triangle

WRITING THE SEGMENT() FUNCTION

Now we need to tell the program how to change a line into a segment that will have different levels. Level 0 will just be a straight line, but the next level will introduce the "bump" in the side. We're really dividing the segment into three segments and then taking the middle segment and replicating it to make it into a little equilateral triangle. We'll change the `snowflake()` function to call another function to draw the segment. This will be the recursive function, because as the levels go up, the segments will become smaller copies of the segment in Figure 10-10.

Figure 10-10: Cutting a segment into thirds and adding a "bump" to the middle third

We'll call the side a *segment*. If the level is 0, the segment is simply a straight line, the side of the triangle. In the next step, a bump is added in the middle of the side. All the segments in Figure 10-10 are the same length, a third of the whole sidelength. This requires 11 steps:

1. Draw a line a third of the sidelength.
2. Translate to the end of the segment you just drew.
3. Rotate –60 degrees (to the left).
4. Draw another segment.
5. Translate to the end of that segment.
6. Rotate 120 degrees (to the right).
7. Draw a third segment.
8. Translate to the end of that segment.
9. Rotate –60 degrees again (to the left).
10. Draw the last segment.
11. Translate to the end of that segment.

Now, instead of drawing a line, the snowflake() function will call a segment() function, which will do the drawing and translating. Add the segment() function in Listing 10-7.

snowflake.pyde

```
def snowflake(sz,level):
    for i in range(3):
        segment(sz,level)
        rotate(radians(120))

def segment(sz,level):
    if level == 0:
        line(0,0,sz,0)
        translate(sz,0)
    else:
        line(0,0,sz/3.0,0)
        translate(sz/3.0,0)
        rotate(radians(-60))
        line(0,0,sz/3.0,0)
        translate(sz/3.0,0)
        rotate(radians(120))
        line(0,0,sz/3.0,0)
        translate(sz/3.0,0)
        rotate(radians(-60))
        line(0,0,sz/3.0,0)
        translate(sz/3.0,0)
```

Listing 10-7: Drawing a "bump" on the sides of the triangle

In the segment() function, if the level is 0, it's just a straight line, and we translate to the end of the line. Otherwise, we have 11 lines of code corresponding to the 11 steps of making a "bump." First, we draw a line a third of the length of the side and then translate to the end of that line. We rotate left (–60 degrees) to draw the second segment in the line. That segment is also a third of the length of the side of the triangle. We translate to the end of that segment and then turn right by rotating 120 degrees. We then draw a segment and turn left one last time by rotating –60 degrees. Finally, we draw a fourth line (segment) and translate to the end of the side.

This draws a triangle if the level is 0 and puts a bump on each side if the level isn't 0. As you can see in Figure 10-8, at every step, every segment in the previous step gets a bump. This would be a headache to do without recursion! But we'll take the line of code that draws a line and change that into a segment, just one level lower. This is the recursive step.

Next, we need to replace each line with a segment one level down, whose length is sz divided by 3. The code for the segment() function is shown in Listing 10-8.

snowflake.pyde

```
def segment(sz,level):
    if level == 0:
        line(0,0,sz,0)
        translate(sz,0)
    else:
        segment(sz/3.0,level-1)
```

```
        rotate(radians(-60))
        segment(sz/3.0,level-1)
        rotate(radians(120))
        segment(sz/3.0,level-1)
        rotate(radians(-60))
        segment(sz/3.0,level-1)
```

Listing 10-8: Replacing the lines with segments

So all we did was replace each instance of `line` in Listing 10-7 (whose level is greater than 0) with `segment()`. Because we don't want to enter an infinite loop, the segments have to be one level down (`level - 1`) from the previous segment. Now we can change the level of the snowflake in the `draw()` function, as shown in the following code, and we'll see different designs, as shown in Figure 10-11.

```
def draw():
    background(255)
    translate(100,height-100)
    snowflake(400,3)
```

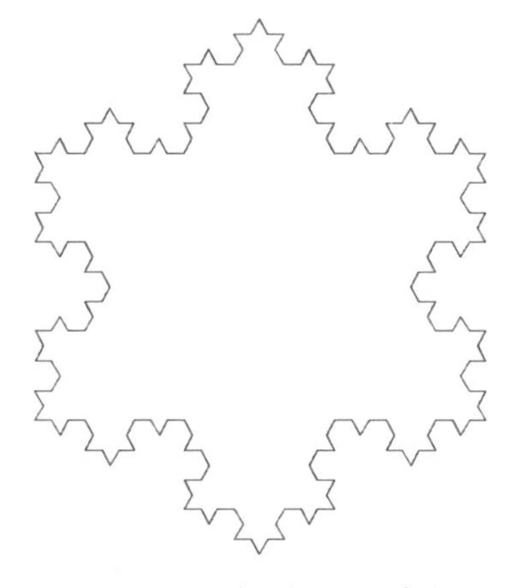

Figure 10-11: A level 3 snowflake

Even better, we can make it interactive by mapping the mouse's x-value to the level. The mouse's x-value can be anywhere from 0 to whatever the width of the screen is. We want to change that range to between 0 and 7. Here's the code for that:

```
level = map(mouseX,0,width,0,7)
```

However, we want only integer levels, so we'll change that value to an integer using `int`, like this:

```
level = int(map(mouseX,0,width,0,7))
```

We'll add that to our draw() function and send the output "level" to the snowflake() function. The entire code for the Koch snowflake is shown in Listing 10-9.

snowflake.pyde

```
def setup():
    size(600,600)

def draw():
    background(255)
    translate(100,200)
    level = int(map(mouseX,0,width,0,7))
    #y(100,level)
    snowflake(400,level)

def snowflake(sz,level):
    for i in range(3):
        segment(sz,level)
        rotate(radians(120))

def segment(sz,level):
    if level == 0:
        line(0,0,sz,0)
        translate(sz,0)
    else:
        segment(sz/3.0,level-1)
        rotate(radians(-60))
        segment(sz/3.0,level-1)
        rotate(radians(120))
        segment(sz/3.0,level-1)
        rotate(radians(-60))
        segment(sz/3.0,level-1)
```

Listing 10-9: Complete code for the Koch snowflake

Now when you run the program and move your mouse left and right, you'll see the snowflake get more "bumps" on its segments, like in Figure 10-12.

Figure 10-12: A level 7 snowflake

How does this help us understand the Coastline Paradox? Looking back at Figure 10-3, let's call the length of the line (the side of the triangle) 1 unit (for example, 1 mile). When we split it in thirds, take out the middle, and add a "bump" two thirds long in the middle, the side is now 1 1/3 units long. It just got 1/3 longer, right? The perimeter of the snowflake (the "coastline") gets 1/3 longer every step. So at the *n*th step, the length of the coastline is $(4/3)^n$ times the perimeter of the original triangle. It might not be possible to see, but after 20 steps, the coastline of the snowflake is so jagged that its total length is over 300 times the original measurement!

SIERPINSKI TRIANGLE

The Sierpinski triangle is a famous fractal first described by Polish mathematician Wacław Sierpiński in 1915, but there are examples of the design on the floors of churches in Italy from as far back as the 11th century! It follows a geometric pattern that's easy to describe, but the design is surprisingly complicated. It works on an interesting recursive idea: draw a triangle for the first level, and for the next level turn each triangle into three smaller triangles at its corners, as shown in Figure 10-13.

Figure 10-13: Sierpinski triangles, levels 0, 1, and 2

The first step is easy: just draw a triangle. Open a new sketch and name it *sierpinski.pyde*. We set it up as usual, with `setup()` and `draw()` functions. In `setup()`, we set the size of the output window to 600 pixels by 600 pixels. In `draw()`, we set the background white and translate to a point (50,450) in the bottom left of the screen to start drawing our triangle. Next, we write a function named `sierpinski()`, similar to what we did with `tree()`, that draws a triangle if the level is 0. The code so far is shown in Listing 10-10.

sierpinski.pyde

```
def setup():
    size(600,600)

def draw():
    background(255)
    translate(50,450)
    sierpinski(400,0)

def sierpinski(sz, level):
    if level == 0: #draw a black triangle
        fill(0)
        triangle(0,0,sz,0,sz/2.0,-sz*sqrt(3)/2.0)
```

Listing 10-10: The setup of the Sierpinski fractal

The sierpinski() function takes two parameters: the size of the figure (sz) and the level variable. The fill color is 0 for black, but you can make it any color you want by using RGB values. The triangle line contains six numbers: the x- and y-coordinates of the three corners of an equilateral triangle with sidelength sz.

As you can see in Figure 10-13, level 1 contains three triangles at each corner of the original triangle. These triangles are also half the size of the triangle in the previous level. What we'll do is create a smaller, lower-level Sierpinski triangle, translate to the next corner, and then rotate 120 degrees. Add the code in Listing 10-11 to the sierpinski() function.

```
def draw():
    background(255)
    translate(50,450)
    sierpinski(400,8)

def sierpinski(sz, level):
    if level == 0: #draw a black triangle
        fill(0)
        triangle(0,0,sz,0,sz/2.0,-sz*sqrt(3)/2.0)
    else: #draw sierpinskis at each vertex
        for i in range(3):
            sierpinski(sz/2.0,level-1)
            translate(sz/2.0,-sz*sqrt(3)/2.0)
            rotate(radians(120))
```

Listing 10-11: Adding the recursive step to the Sierpinski program

This new code tells Processing what to do when the level isn't 0 (the line for i in range(3): means "repeat this three times"): draw a half-sized Sierpinski triangle of one level lower, and then translate halfway across and halfway up the equilateral triangle and turn right 120 degrees. Notice the sierpinski() function in sierpinski(sz/2.0,level-1) is executed inside the definition of the sierpinski() function itself. That's the recursive step! When you call

```
sierpinski(400,8)
```

in the draw() function, you get a level 8 Sierpinski triangle, which you see in Figure 10-14.

An interesting thing about the Sierpinski triangle is that it shows up in other fractals too, like the next one, which doesn't start with a triangle.

Figure 10-14: A level 8 Sierpinski triangle

SQUARE FRACTAL

We can make the Sierpinski triangle out of squares too. For example, we can create a square, remove the lower-right quadrant, and then replace each remaining quadrant with the resulting shape. When we repeat this process, we should get something like Figure 10-15.

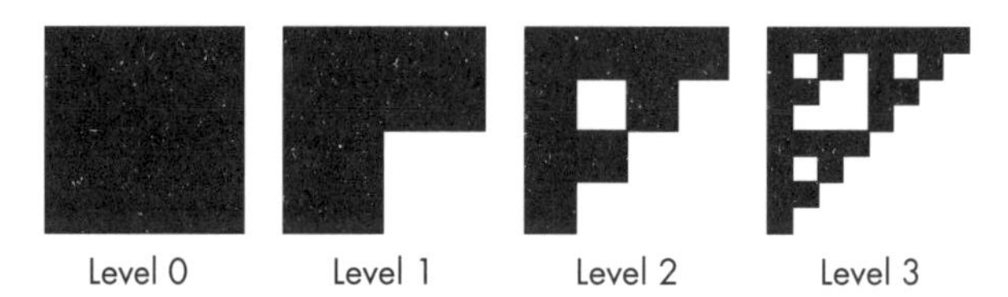

Figure 10-15: The square fractal at levels 0, 1, 2, and 3

To create this fractal, we have to make each of the three smaller squares into a copy of the whole. Start a new Processing sketch called *squareFractal.pyde* and then set up the sketch with the code in Listing 10-12.

squareFractal.pyde

```
def setup():
    size(600,600)
    fill(150,0,150) #purple
    noStroke()

def draw():
    background(255)
    translate(50,50)
    squareFractal(500,0)

def squareFractal(sz,level):
    if level == 0:
        rect(0,0,sz,sz)
```

Listing 10-12: Creating the `squareFractal()` function

We can use the RGB values for purple in the `setup()` function just because we won't be changing the fill anywhere else. We use `noStroke()` so that we won't see black outlines on the squares. In the `draw()` function we call the `squareFractal()` function, telling it to make the size of each square 500 pixels and level 0. In the function definition, we tell the program to simply draw a square if the level is zero. This should give us a nice big purple square, as shown in Figure 10-16.

Figure 10-16: Purple square (level 0)

For the next level, we'll make squares of half the sidelength of the initial square. One will be positioned at the top left of the figure; then we'll translate around to put the other two squares at the bottom left and top right of Figure 10-16. Listing 10-13 does this while leaving out a quarter of the big square.

squareFractal.pyde

```
def squareFractal(sz,level):
    if level == 0:
        rect(0,0,sz,sz)
    else:
        rect(0,0,sz/2.0,sz/2.0)
        translate(sz/2.0,0)
        rect(0,0,sz/2.0,sz/2.0)
        translate(-sz/2.0,sz/2.0)
        rect(0,0,sz/2.0,sz/2.0)
```

Listing 10-13: Adding more squares to the square fractal

Here, we draw a big square if the level is 0. If the level is not 0, we add a smaller square in the top left of the screen, translate to the right, add another smaller square in the top right, translate left (negative x) and down (positive y), and add a smaller square at the bottom left of the screen.

That's the next level, and when we update `squareFractal(500,0)` in the `draw()` function to `squareFractal(500,1)`, it should give us a square with the bottom-right quarter left out, as shown in Figure 10-17.

Figure 10-17: The next level of the square fractal

For the next levels, we want each of the squares to be further subdivided into fractals, so we'll replace the `rect` lines with `squareFractal()`, divide the value in `sz` by 2, and tell it to move one level down, like in Listing 10-14.

squareFractal.pyde

```
def squareFractal(sz,level):
    if level == 0:
        rect(0,0,sz,sz)
    else:
        squareFractal(sz/2.0,level-1)
        translate(sz/2.0,0)
        squareFractal(sz/2.0,level-1)
```

```
        translate(-sz/2.0,sz/2.0)
        squareFractal(sz/2.0,level-1)
```

Listing 10-14: Adding the recursive step to the square fractal

In Listing 10-14, notice that the rect lines (when the level isn't 0) are replaced with squareFractal(). When we call squareFractal(500,2) in the draw() function, we don't get the output we were expecting—we get Figure 10-18 instead.

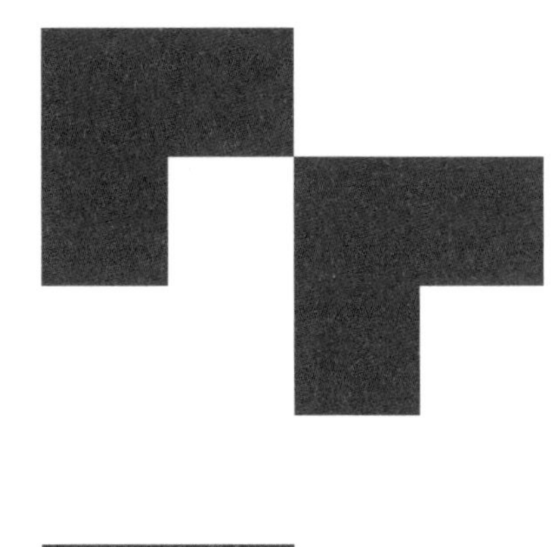

Figure 10-18: Not what we were expecting!

This is because we didn't translate back to the starting point like we did with our Y fractal earlier in the chapter.

Although we can calculate how much to translate manually, we can also use the pushMatrix() and popMatrix() functions in Processing, which you learned about in Chapter 5.

We can use the pushMatrix() function to save the current orientation of the screen—that is, where the origin (0,0) is located and how much the grid is rotated. After that, we can do as much translating and rotating as we like and then use the popMatrix() function to return to the saved orientation without any calculating!

Let's add pushMatrix() at the beginning of the squareFractal() function and popMatrix() at the end, like in Listing 10-15.

squareFractal.pyde

```
def squareFractal(sz,level):
    if level == 0:
        rect(0,0,sz,sz)
    else:
        pushMatrix()
        squareFractal(sz/2.0,level-1)
        translate(sz/2.0,0)
        squareFractal(sz/2.0,level-1)
        translate(-sz/2.0,sz/2.0)
        squareFractal(sz/2.0,level-1)
        popMatrix()
```

Listing 10-15: Using pushMatrix() and popMatrix() to complete the squares

Now, each of the smaller squares from level 1 should be transformed into a fractal, with the bottom-right square removed, as shown in Figure 10-19.

Figure 10-19: Level 2 of the square fractal

Now let's try making our mouse generate the level numbers like we've done before by replacing `squareFractal(500,2)` with the code in Listing 10-16.

squareFractal.pyde

```
def draw():
    background(255)
    translate(50,50)
    level = int(map(mouseX,0,width,0,7))
    squareFractal(500,level)
```

Listing 10-16: Making the square fractal interactive

At higher levels, the square fractal looks a lot like the Sierpinski triangle, as you can see in Figure 10-20!

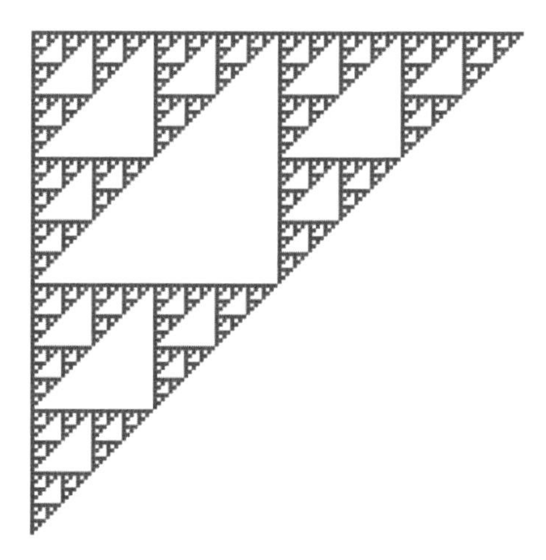

Figure 10-20: High-level square fractals look like the Sierpinski triangle!

DRAGON CURVE

The final fractal we'll create looks different from the others we've created so far in that the shapes on each level don't get smaller, they get bigger. Figure 10-21 shows an example of the dragon curve for levels 0 through 3.

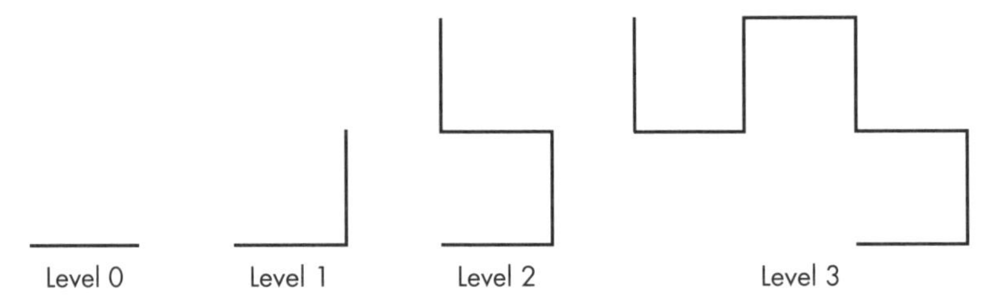

Figure 10-21: The first four levels of the dragon curve

As mathematical entertainer Vi Hart shows in one of her YouTube videos, the second half of the dragon curve is a perfect copy of the first half, and she models it by folding and then unfolding pieces of paper. The third level (level 2) in Figure 10-21 looks like two left turns followed by a right turn. The "hinge" or "fold" is at the midpoint of each dragon curve. See if you can find it in your dragon curves! Later, you'll rotate part of the curve dynamically to match the next-level curve.

Open a new Processing sketch and name it *dragonCurve.pyde*. To create this fractal, we first create a function for the "left dragon," as in Listing 10-17.

dragonCurve.pyde

```
def setup():
    size(600,600)
    strokeWeight(2) #a little thicker lines

def draw():
    background(255)
    translate(width/2,height/2)
    leftDragon(5,11)

def leftDragon(sz,level):
    if level == 0:
        line(0,0,sz,0)
        translate(sz,0)
    else:
        leftDragon(sz,level-1)
        rotate(radians(-90))
        rightDragon(sz,level-1)
```

Listing 10-17: Writing the `leftDragon()` function

After the usual `setup()` and `draw()` functions, we define our `leftDragon()` function. If the level is 0, we just draw a line and then translate along the line. It's kind of like the turtle from Chapter 1 drawing a line as it walks along. If the level is greater than 0, make a left dragon (one level down), turn left 90 degrees, and make a right dragon (one level down).

Now we'll make the "right dragon" function (see Listing 10-18). It's pretty similar to the `leftDragon()` function. If the level is 0, simply draw a line and move along it. Otherwise, make a left dragon, and this time turn *right* 90 degrees and make a right dragon.

dragonCurve.pyde

```
def rightDragon(sz,level):
    if level == 0:
        line(0,0,sz,0)
        translate(sz,0)
    else:
        leftDragon(sz,level-1)
        rotate(radians(90))
        rightDragon(sz,level-1)
```

Listing 10-18: Writing the `rightDragon()` function

It's interesting that the recursive statement in this case is not only inside one function, but it also jumps back and forth from the left dragon function to the right dragon function! Execute it, and the 11th level will look like Figure 10-22.

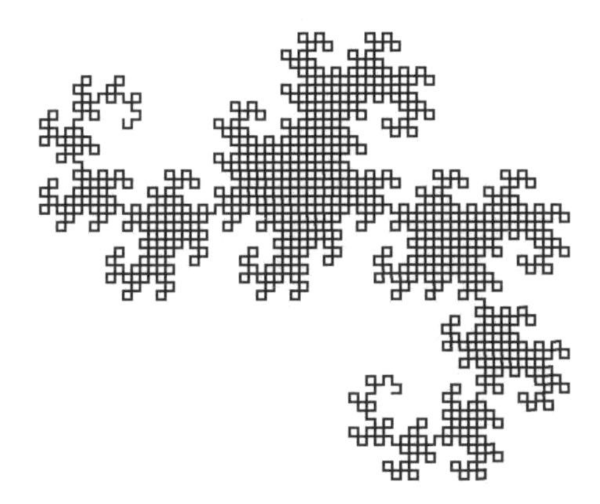

Figure 10-22: A level 11 dragon curve

Far from being simply a chaotic jumble of angles, this fractal starts to look like a dragon after enough levels! Remember I said the dragon curve is "folded" in the middle? In the version shown in Listing 10-19, I've added a few variables to change the level and the size, and I made an `angle` variable change with the mouse's x-coordinate. This will rotate a dragon curve around a "hinge" in the middle of the next-level dragon curve. See how you can simply rotate the curve to get both halves of the next-level curve!

dragonCurve.pyde

```
❶ RED = color(255,0,0)
  BLACK = color(0)
```

```
def setup():
❷ global thelevel,size1
   size(600,600)
❸ thelevel = 1
   size1 = 40

def draw():
    global thelevel
    background(255)
    translate(width/2,height/2)
❹ angle = map(mouseX,0,width,0,2*PI)
    stroke(RED)
    strokeWeight(3)
    pushMatrix()
    leftDragon(size1,thelevel)
    popMatrix()
    leftDragon(size1,thelevel-1)
❺ rotate(angle)
    stroke(BLACK)
    rightDragon(size1,thelevel-1)

def leftDragon(sz,level):
    if level == 0:
        line(0,0,sz,0)
        translate(sz,0)
    else:
        leftDragon(sz,level-1)
        rotate(radians(-90))
        rightDragon(sz,level-1)

def rightDragon(sz,level):
    if level == 0:
        line(0,0,sz,0)
        translate(sz,0)
    else:
        leftDragon(sz,level-1)
        rotate(radians(90))
        rightDragon(sz,level-1)

def keyPressed():
    global thelevel,size1
❻ if key == CODED:
        if keyCode == UP:
            thelevel += 1
        if keyCode == DOWN:
            thelevel -= 1
        if keyCode == LEFT:
            size1 -= 5
        if keyCode == RIGHT:
            size1 += 5
```

Listing 10-19: A dynamic dragon curve

In Listing 10-19, we add a couple of colors ❶ to use for the curves. In the `setup()` function, we declare two global variables, `thelevel` and `size1` ❷, whose initial values we declare at ❸ and which we change with the arrow keys in the `keyPressed()` function at the end of the file.

In the `draw()` function, we link an `angle` variable ❹ to the x-position of the mouse. After that, we set the stroke color to red, make the stroke weight a little heavier, and draw a left dragon with the initial values of `thelevel` and `size1`. The `pushMatrix()` and `popMatrix()` functions, as you'll remember, simply return the drawing point to the original spot, to draw another curve. Then we rotate the grid by however many radians the angle variable is ❺, and draw another dragon curve, in black. The `leftDragon()` and `rightDragon()` functions are exactly the same as before.

Processing's built-in `keyPressed()` function could come in handy for changing variables in a sketch! All you have to do is declare the global variables you want to change with the left (in this case), right, up, and down arrow keys on the keyboard. Note that `CODED` ❻ just means it's not a letter or character key. Finally, it checks which arrow key is being pressed and makes the level variable go up or down (if the up or down arrow key is being pressed) or the size variable go up or down (if the left or right arrow key is being pressed).

When you run this version of the *dragonCurve* sketch, it draws a dragon curve at level 5 in red; then you can rotate a level 4 curve and see how the level 5 curve is made up of two level 4's, just rotated in the middle, as shown in Figure 10-23.

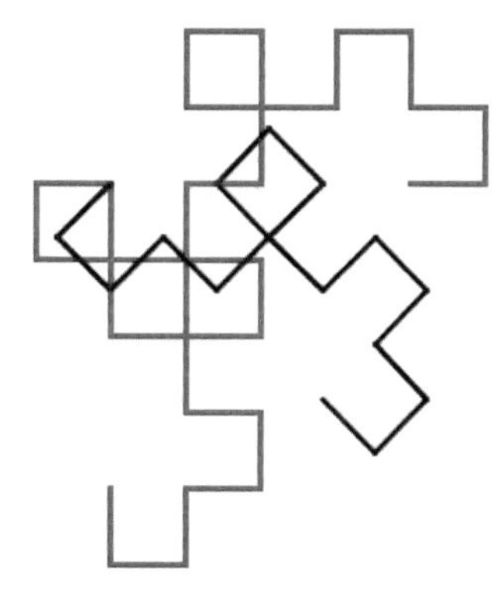

Figure 10-23: A level 5 dragon curve and a dynamic, interactive level 4 curve

When you move the mouse, the black dragon curve should rotate, and you can see how it fits both halves of the red curve. The up and down arrow keys control the level of the curve; press the up arrow key and the curve gets longer. If the curve extends off the display window, use the left arrow key to make each segment shorter, so it'll fit on the screen. The right arrow key makes it bigger.

This makes sense, because the `leftDragon()` function comes first, turns left, and makes a right dragon curve. The `rightDragon()` function just turns the opposite way from `leftDragon()`: it makes a right turn in the middle instead of a left. No wonder it turns out to be a perfect copy.

SUMMARY

We've only scratched the surface of fractals, but hopefully you got a taste of how beautiful fractals can be and how powerful they are at modeling the messiness of nature. Fractals and recursion can help us reevaluate our ideas about logic and measurement. The question is no longer "how long is the coastline?" but rather "how jagged is it?"

For fractal lines like coastlines and meandering rivers, the standard characteristic is the scale of self-similarity, or how much we have to scale the map up by before it looks like a different scale of the same thing. This is effectively what you did by feeding `0.8*sz`, `sz/2.0`, or `sz/3.0` into the next level.

In the next chapter, we'll create cellular automata (CAs), which we'll draw as little squares on the screen that are born, grow, and respond to their surroundings. Just like with our grass-eating sheep in Chapter 9, we'll create CAs and let them run—and just like with fractals, we'll watch the surprising and beautiful patterns that are created from very simple rules.

11

CELLULAR AUTOMATA

I like to put a humidifier and a dehumidifier in a room and just let them fight it out.
—Steven Wright

Math equations are a very powerful tool for modeling things we can measure; equations even got us to the moon, after all. But as powerful as they are, equations are of limited use in the biological and social sciences because organisms don't grow according to equations.

Organisms grow in an environment among many other organisms and spend their day performing innumerable interactions. That web of interactions determines how something will grow, and equations often can't capture this complicated relationship. Equations can help us calculate the energy or mass converted by a single interaction or reaction, but to model a biological system, for example, you'd have to repeat that calculation hundreds or thousands of times.

Fortunately, there's a tool that models how cells, organisms, and other living systems grow and change according to their environment. Because of their similarity to independent biological organisms, these models are called *cellular automata (CAs)*. The term *automaton* refers to something that can run on its own. Figure 11-1 shows two examples of cellular automata generated using a computer.

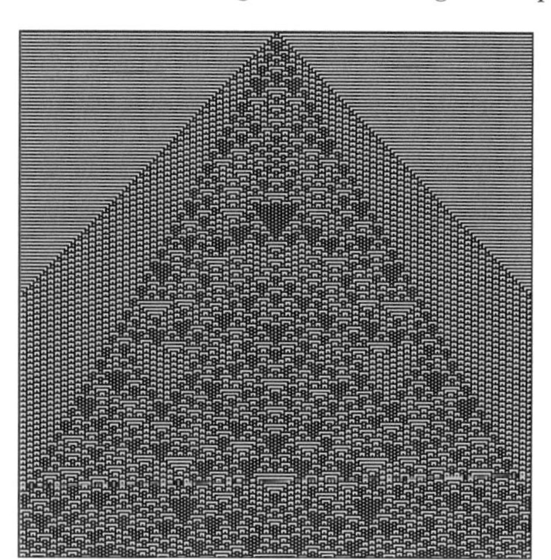
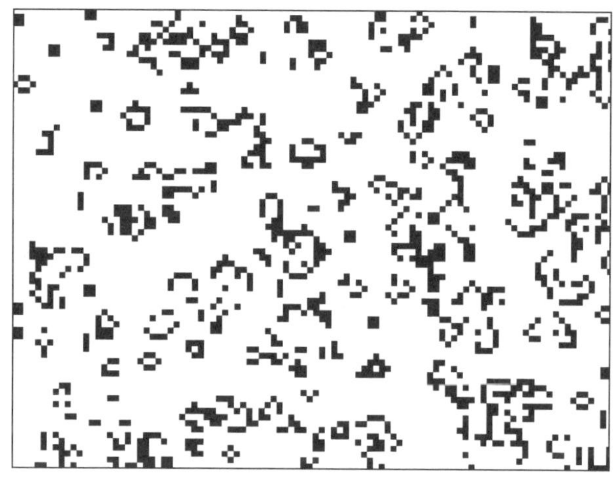

Figure 11-1: An example of an elementary cellular automaton, and a screen full of virtual organisms

The CAs we'll create in this chapter are grids made up of *cells*. Each cell in a CA has a number of *states* (for example, on/off, alive/dead, or colored/blank). Cells change according to the state of their neighbors, which allows them to grow and change as if they were alive!

CAs have been the subject of some study, as far back as the 1940s, but they really took off when computers became more commonplace. In fact, CAs can really only be studied using computers because, even though they follow very simple rules, like "if an organism doesn't have enough neighbors, it dies," these rules produce useful results only if hundreds or thousands of these organisms are created and allowed to run for hundreds or thousands of generations.

Because math is the study of patterns, the math topic of cellular automata is rife with interesting ideas, programming challenges, and endless possibilities for beautiful output!

CREATING A CELLULAR AUTOMATON

Open a new Processing sketch and name it *cellularAutomata.pyde*. Let's start with a square grid where our cells will reside. We can easily draw a 10-by-10 grid of squares of size 20, as shown in Listing 11-1.

cellularAutomata.pyde

```
def setup():
    size(600,600)

def draw():
    for x in range(10):
```

```
        for y in range(10):
            rect(20*x,20*y,20,20)
```

Listing 11-1: Creating a grid of squares

Save and run this sketch, and you should see a grid like the one shown in Figure 11-2.

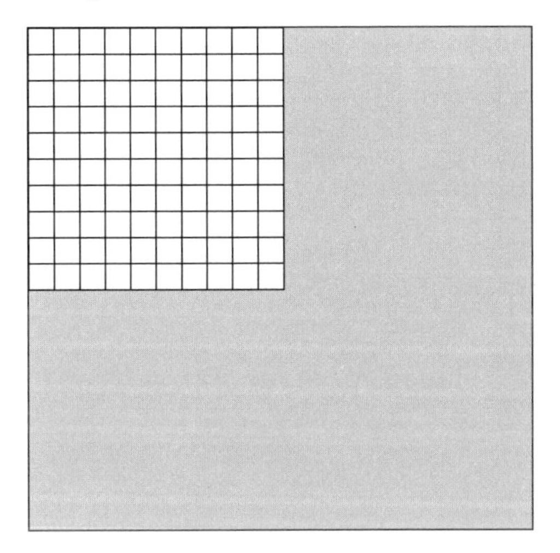

Figure 11-2: A 10 × 10 grid

However, we need to change a bunch of numbers every time we want bigger cells, for example, or a grid with different dimensions. Therefore, it's much easier to change things later if we use variables. Because the keywords `height`, `width`, and `size` already exist for the graphics window, we have to use different variable names. Listing 11-2 improves on Listing 11-1 by creating a grid that's easy to resize, with cells that are also easy to resize—all by using variables.

cellularAutomata.pyde

```
GRID_W = 15
GRID_H = 15

#size of cell
SZ = 18
def setup():
    size(600,600)

def draw():
    for c in range(GRID_W): #the columns
        for r in range(GRID_H): #the rows
            rect(SZ*c,SZ*r,SZ,SZ)
```

Listing 11-2: Improved grid program using variables

We create variables for the height (`GRID_H`) and width (`GRID_W`) of the grid using all capital letters to indicate that these are constants and their values won't be changing. The size of the cell is also a constant (for now), so we capitalize it as well (SZ) when declaring its initial value. Now when you run this code, you should see a larger grid, like the one shown in Figure 11-3.

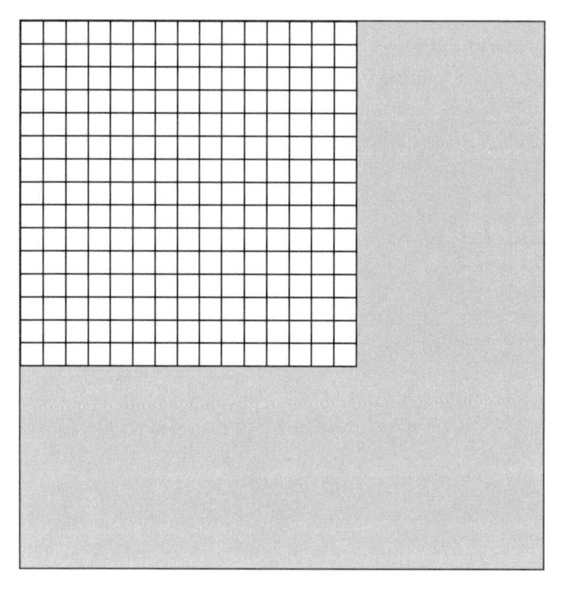

Figure 11-3: A larger grid, made with variables

WRITING A CELL CLASS

We need to write a class because every cell we create needs its own location, state ("on" or "off"), neighbors (the cells next to it), and so on. We create the Cell class by adding the code shown in Listing 11-3.

cellularAutomata.pyde

```
#size of cell
SZ = 18

class Cell:
    def __init__(self,c,r,state=0):
        self.c = c
        self.r = r
        self.state = state

    def display(self):
        if self.state == 1:
            fill(0) #black
        else:
            fill(255) #white
        rect(SZ*self.r,SZ*self.c,SZ,SZ)
```

Listing 11-3: Creating the Cell class

The cell's initial state property is 0 (or off). The code state=0 in the parameters of the __init__ method means that if we don't specify a state, state is set to 0. The display() method just tells the Cell object how to display itself on the screen. If it's "on," the cell is black; otherwise, it's white. Also, each cell is a square, and we need to spread out the cells by multiplying their column and row numbers by their size (self.SZ).

After the draw() function, we need to write a function to create an empty list to put our Cell objects in and use a nested loop to append these Cell objects to the list instead of drawing them one by one, as shown in Listing 11-4.

cellular Automata.pyde

```
def createCellList():
    '''Creates a big list of off cells with
    one on Cell in the center'''
❶   newList=[]#empty list for cells
    #populate the initial cell list
    for j in range(GRID_H):
      ❷ newList.append([]) #add empty row
        for i in range(GRID_W):
          ❸ newList [j].append(Cell(i,j,0)) #add off Cells or zeroes
    #center cell is set to on
❹   newList [GRID_H//2][GRID_W//2].state = 1
    return newList
```

Listing 11-4: Function for creating a list of cells

First, we create an empty list called newList ❶ and add an empty list as a row ❷ to be filled in with Cell objects ❸. Then, we get the index of the center square by dividing the number of rows and columns by 2 (the double slash means integer division) and setting that cell's state property to 1 (or "on") ❹.

In setup(), we'll use the createCellList() function and declare cellList as a global variable so it can be used in the draw() function. Finally, in draw(), we'll loop over each row in cellList and update it. The new setup() and draw() functions are shown in Listing 11-5.

```
def setup():
    global cellList
    size(600,600)
    cellList = createCellList()

def draw():
    for row in cellList:
        for cell in row:
            cell.display()
```

Listing 11-5: The new setup() and draw() functions for creating a grid

However, when we run this code, we get a grid with smaller cells in the corner of the display window, as shown in Figure 11-4.

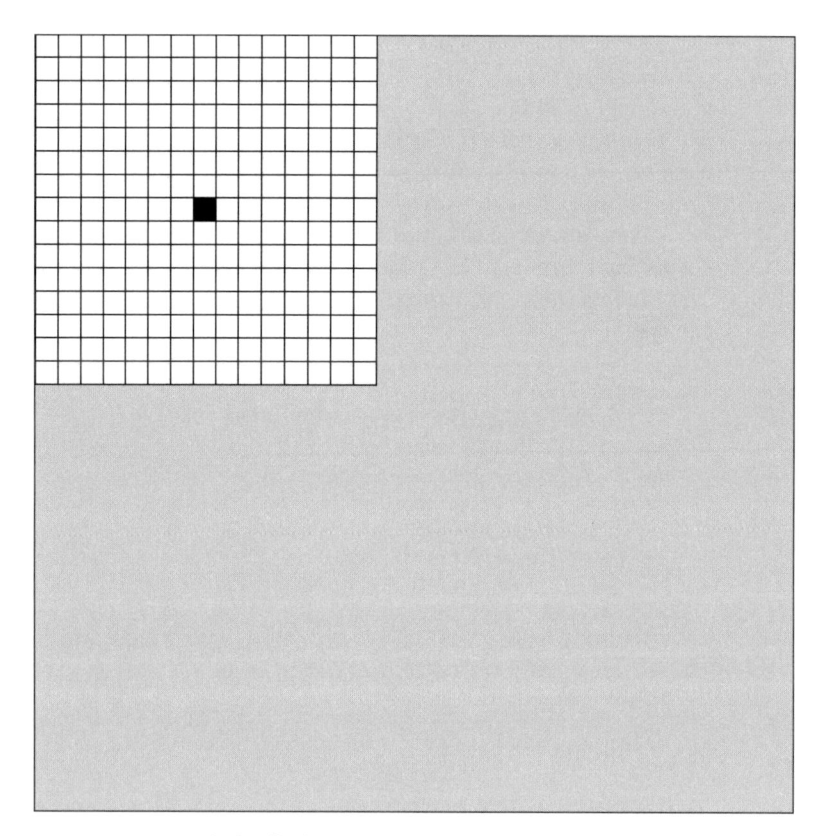

Figure 11-4: A grid of cells that's not yet centered

Now we're able to make as big or small a list of cells as we want by changing the size of our 15-by-15 grid.

RESIZING EACH CELL

To resize our cells, we can make `SZ` automatically dependent on the width of the window. Right now the width is 600, so let's change `setup()` using the code in Listing 11-6.

cellularAutomata.pyde

```
def setup():
    global SZ,cellList
    size(600,600)
    SZ = width // GRID_W + 1
    cellList = createCellList()
```

Listing 11-6: Resizing the cells to autofit the display window

The double forward slash (`//`) means *integer division*, which returns only the integer part of the quotient. Now, when you run the program, it should produce a grid with all empty cells except for one colored cell in the center, like in Figure 11-5.

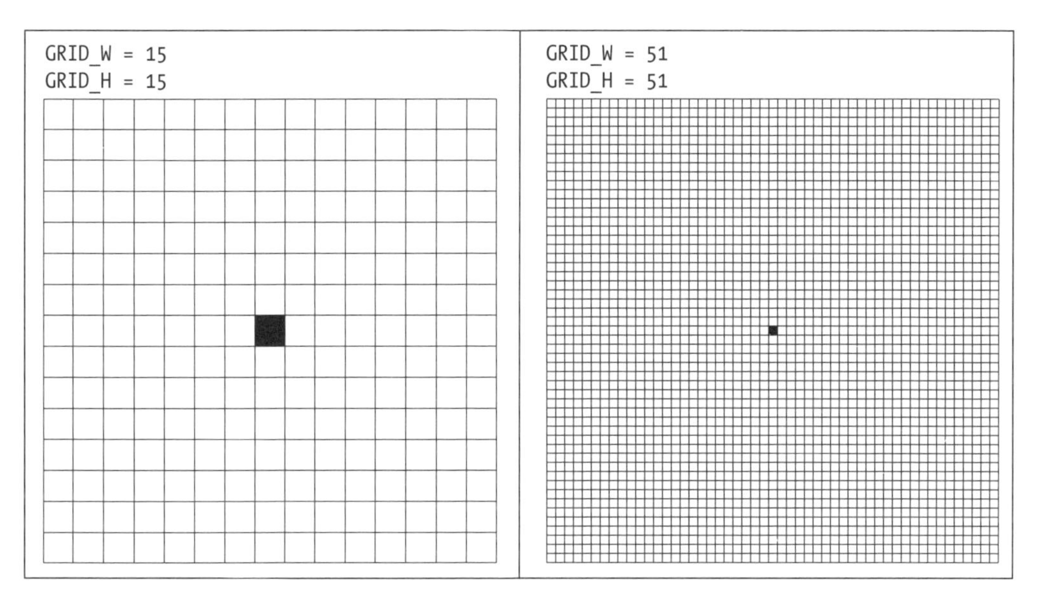

Figure 11-5: Grids with the center cell "on"

Note that this code works better when you add 1 to `SZ`, the size of the `Cell`, as in Listing 11-16, because otherwise the grid sometimes doesn't fill the whole display window. But feel free to leave it out.

MAKING A CA GROW

Now we want to make the cells change according to the number of their neighbors whose state is "on." This section was inspired by a two-dimensional CA from Stephen Wolfram's *New Kind of Science*. You can see how a version of this CA grows in Figure 11-6.

Figure 11-6: Stages of growth of a cellular automaton

In this design, if a cell has *one or four* neighbors that are on, we make it turn on (and stay on).

PUTTING THE CELLS INTO A MATRIX

It's easy to find the cells immediately before and after a cell in the list, which gives us the neighbors to its left and right. But how do we find the neighbors above and below a cell? To do this more easily, we can put the cells in a two-dimensional *array* or *matrix,* which is a list with lists for the rows. That way, if a cell is in column 5, for example, we know that its "above" and "below" neighbors will also be in column 5.

In the `Cell` class, we add a method called `checkNeighbors()` so that a cell can count how many of its neighbors are on, and if the count is 1 or 4, that cell will return 1 for "on." Otherwise, it returns 0 for "off." We begin by checking the neighbor above:

```
def checkNeighbors(self):
    if self.state == 1: return 1 #on Cells stay on
    neighbs = 0
    #check the neighbor above
    if cellList[self.r-1][self.c].state == 1:
        neighbs += 1
```

This code checks for the item in `cellList` that's in the same column (`self.c`) but in the previous row (`self.r - 1`). If that item's `state` property is 1, then it's on, and we increment the `neighbs` variable by 1. Then we have to do the same for the cell's neighbor below, and then for the neighbors to the left and right. Do you see an easy pattern here?

```
cellList[self.r - 1][self.c + 0] #above
cellList[self.r + 1][self.c + 0] #below
cellList[self.r + 0][self.c - 1] #left
cellList[self.r + 0][self.c + 1] #right
```

We only need to keep track of the change in the row number and the change in the column number. There are only four directions we need to check, for the "one to the left, one to the right" neighbors, and so on: `[-1,0]`, `[1,0]`, `[0,-1]` and `[0,1]`. If we call those `dr` and `dc` (*d*, or the Greek letter *delta,* is the traditional math symbol for change), we can keep from repeating ourselves:

cellularAutomata.pyde

```
def checkNeighbors(self):
    if self.state == 1: return 1 #on Cells stay on
    neighbs = 0  #check the neighbors
    for dr,dc in [[-1,0],[1,0],[0,-1],[0,1]]:
        if cellList[self.r + dr][self.c + dc].state == 1:
            neighbs += 1
    if neighbs in [1,4]:
        return 1
    else:
        return 0
```

Finally, if the neighbor count is 1 or 4, the `state` property will be set to 1. In Python, `if neighbs in [1,4]` is the same as saying `if neighbs == 1 or neighbs == 4:`.

CREATING THE CELL LIST

So far, we've created the cell list by running the `createCellList()` function in `setup()` and assigning the output to `cellList`, and we've gone through every row in `cellList` and updated each cell in the row. Now we have to check whether the rules work. The four squares surrounding the center cell should change state in the next step. That means we'll have to run the `checkNeighbors()` method and then show the result. Update your `draw()` function as follows:

```
def draw():
    for row in cellList:
        for cell in row:
          ❶ cell.state = cell.checkNeighbors()
            cell.display()
```

The updated line ❶ runs all the `checkNeighbors()` code and sets the cell on or off according to the result. Run it, and you should get the following error:

```
IndexError: index out of range: 15
```

The error is in the line that checks the neighbor to the right. Sure enough, because there are only 15 cells in a row, it makes sense that the 15th cell has no neighbor to the right.

If a cell has no neighbor to the right (meaning its column number is `GRID_W` minus one), we obviously don't need to check that neighbor and can continue on to the next cell. The same for checking the neighbor above the cells in row 0, because they have no cells above them. Similarly, the cells in column 0 have no neighbors to the left, and the cells in row 14 (`GRID_H` minus 1) have no cells below them. In Listing 11-7, we add a valuable Python trick called *exception handling* to the `checkNeighbors()` method using the keywords `try` and `except`.

cellular Automata.pyde

```
    def checkNeighbors(self,cellList):
        if self.state == 1: return 1 #on Cells stay on
        neighbs = 0
        #check the neighbors
        for dr,dc in [[-1,0],[1,0],[0,-1],[0,1]]:
          ❶ try:
                if cellList[self.r + dr][self.c + dc].state == 1:
                    neighbs += 1
          ❷ except IndexError:
                continue
        if neighbs in [1,4]:
            return 1
        else:
            return 0
```

Listing 11-7: Adding conditionals to `checkNeighbors()`

The try keyword ❶ literally means "try to run this next line of code." In the earlier error message, we got an IndexError. We use the except keyword ❷ to mean "if you get this error, do this." Therefore, if we get an IndexError, we continue on to the next loop. Run this code, and you'll get something interesting, as shown in Figure 11-7. This is definitely not what we saw in Figure 11-6.

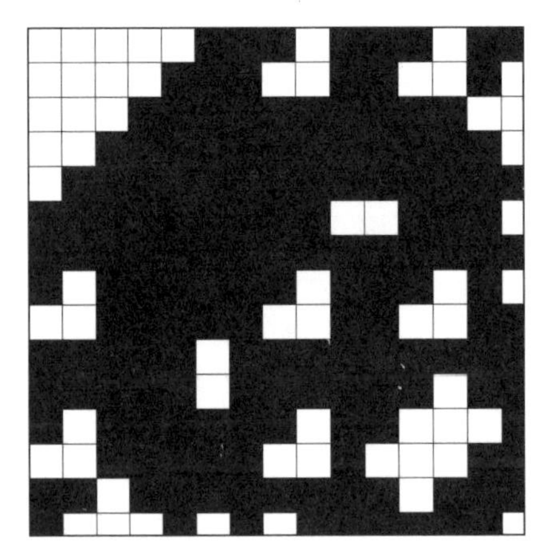

Figure 11-7: Not what we expected!

The problem is that we're checking neighbors and changing the state of the current cell. Then the cell's neighbors are checking their neighbors, but they're checking the new state of their neighbors. We want all the cells to check their neighbors and save the information in a new list; then, when all the cells are done, we can update the grid all at once. That calls for another list for our cells, newList, that will replace cellList at the end of the loop.

So all we need to do is declare that newList is equal to cellList, right?

```
cellList = newList  #?
```

Although that seems to make sense, Python doesn't copy the contents of newList over the previous contents of cellList, which is what you might have expected. It technically refers to the newList, but when you change newList, you end up changing cellList as well.

PYTHON LISTS ARE STRANGE

Python lists have an odd behavior. Let's say you declare a list and set another one equal to it, and then you change the first list. You wouldn't expect the second one to change too, but that's exactly what happens, as shown here:

```
>>> a = [1,2,3]
>>> b = a
>>> b
[1, 2, 3]
```

```
>>> a.append(4)
>>> a
[1, 2, 3, 4]
>>> b
[1, 2, 3, 4]
```

As you can see, we created list a, then assigned the value of list a to list b. When we change list a without updating list b, Python also changes list b!

LIST INDEX NOTATION

One way to make sure when we're updating one list that we're not updating another one accidentally is to use index notation. Giving list b all the contents of list a should prevent this from happening:

```
>>> a = [1,2,3]
>>> b = a[::]
>>> b
[1, 2, 3]
>>> a.append(4)
>>> a
[1, 2, 3, 4]
>>> b
[1, 2, 3]
```

Here, we use `b = a[::]` to say "assign all the contents inside list a to the variable b," as opposed to simply declaring that list a is equal to list b. This way, the lists aren't linked to each other.

After we declare SZ, we need to add the following line of code to declare the initial value of the generation variable, which will keep track of which generation we're looking at:

```
generation = 0
```

We're going to avoid the list reference problem by using the index notation at the end of the updating code. Let's create a new update() function after draw() so that all the updating will be done in that separate function. Listing 11-8 shows how your setup() and draw() functions should look.

cellular Automata.pyde

```
def setup():
    global SZ, cellList
    size(600,600)
    SZ = width // GRID_W + 1
    cellList = createCellList()

def draw():
    global generation,cellList
    cellList = update(cellList)
    for row in cellList:
        for cell in row:
            cell.display()
```

```
    generation += 1
    if generation == 3:
        noLoop()

def update(cellList):
    newList = []
    for r,row in enumerate(cellList):
        newList.append([])
        for c,cell in enumerate(row):
            newList[r].append(Cell(c,r,cell.checkNeighbors()))
    return newList[::]
```

Listing 11-8: Checking whether the updating is working and then stopping after three generations

We create the first `cellList` once in the `setup()` function and then declare it a global variable so we can use it in other functions. In the `draw()` function, we use the `generation` variable for however many generations we want to check (in this case, three); then we make a call to update the `cellList`. We draw the cells as before, using the `display()` method, and then increment `generation` and check whether it has reached our desired generation. If it has, the built-in Processing function `noLoop()` stops the loop.

We use `noLoop()` to turn off the infinite loop, because we only want to draw the given number of generations. If you comment it out, the program will keep going! Figure 11-8 shows what the CA looks like after three generations.

Figure 11-8: A working CA!

What's great about using variables for our grid size is that we can change the CA drastically by simply changing the GRID_W and GRID_H variables, like so:

```
GRID_W = 41
GRID_H = 41
```

If we increase the number of generations to 13 (in the line that currently reads if generation == 3), the output should look like Figure 11-9.

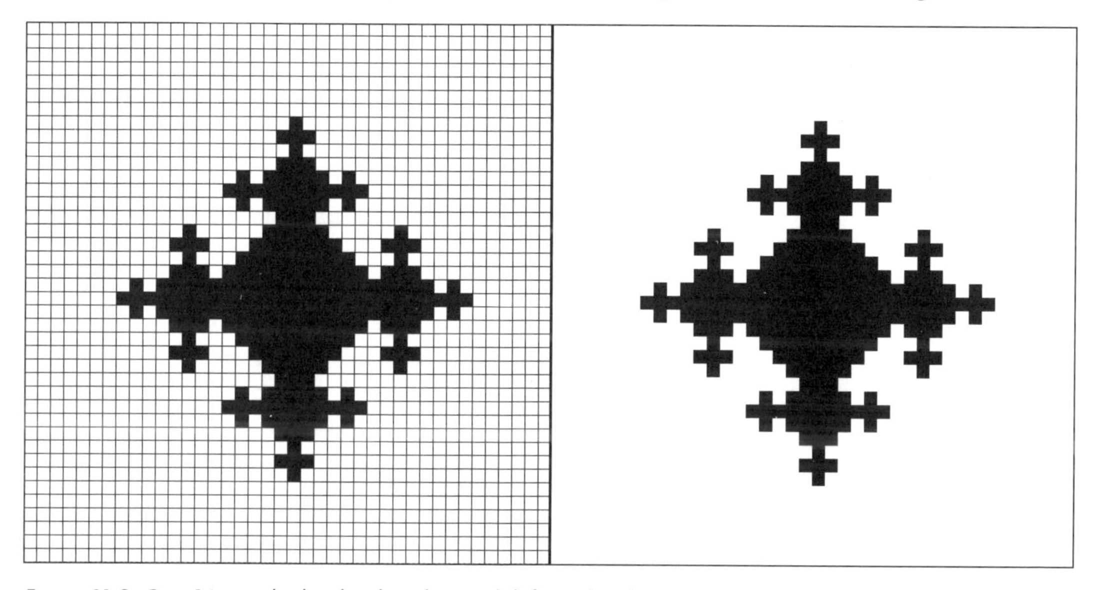

Figure 11-9: Our CA at a higher level, with a grid (left) and without a grid (right)

To remove the grid around the empty cells in the CA, simply add this line to the setup() function:

```
    noStroke()
```

That should turn off the outline around the squares, but the fill color will still be drawn, like Figure 11-9.

So far we've done a lot! We've created two-dimensional lists, filled them with cells, and turned on certain cells according to a simple rule. Then we updated the cells and displayed them. The CA just keeps growing!

EXERCISE 11-1: MANUALLY GROWING THE CA

Use the keyPressed() function you learned about in Chapter 10 to manually grow the CA.

LETTING YOUR CA GROW AUTOMATICALLY

If you want the CA to cycle from level 0 to a maximum number of generations (you choose the right number for your window), simply change the draw() function to what's shown in Listing 11-9.

cellular Automata.pyde

```
def draw():
    global generation,cellList
❶   frameRate(10)
    cellList = update(cellList)
    for row in cellList:
        for cell in row:
            cell.display()
    generation += 1
❷   if generation == 30:
        generation = 1
        cellList = createCellList()
```

Listing 11-9: Making the CA grow and regrow automatically

To slow down the animation, we use Processing's built-in frameRate() function ❶. The default is 60 frames per second, so here we slowed it down to 10. Then we tell the program that if the generation variable reaches 30 ❷ (you can change this to another number), reset generation to 1, and create a new cellList. Now you should be able to watch the CA grow as quickly or slowly as you want. Change the rule and see how that changes the CA. You can change the colors too!

We've just taken a simple rule (if a cell has 1 or 4 neighbors, it's "on") and wrote a program to apply that rule to thousands of cells at once! The result looks like a living, growing organism. Now we'll expand our code into a famous CA where the virtual organisms can move around, grow, and die!

PLAYING THE GAME OF LIFE

In a 1970 issue of *Scientific American*, math popularizer Martin Gardner brought attention to a strange and wonderful game where cells live or die according to how many neighbors they have. The brainchild of English mathematician John Conway, this game features three simple rules:

1. If a living cell has less than two living neighbors, it dies.
2. If a living cell has more than three living neighbors, it dies.
3. If a dead cell has exactly three living neighbors, it comes to life.

With a simple set of rules like that, it's surprising how intricate this game gets. In 1970, most people could only use checkers on a board to visualize the game, and one generation could take quite a while to calculate. Conveniently, we have a computer, and the CA code we just wrote has most of the code necessary to create this game in Python. Save the CA file we've been working on so far and then save it with different name, like *GameOfLife*.

In this game, our cells will have diagonal neighbors too. That means we have to add four more values to our dr,dc line. Listing 11-10 shows the changes you need to make to the checkNeighbors() code.

GameOfLife.pyde

```
def checkNeighbors(self):
    neighbs = 0  #check the neighbors

❶   for dr,dc in [[-1,-1],[-1,0],[-1,1],[1,0],[1,-1],[1,1],[0,-1],[0,1]]:
        try:
            if cellList[self.r + dr][self.c + dc].state == 1:
                neighbs += 1
        except IndexError:
            continue
❷   if self.state == 1:
        if neighbs in [2,3]:
            return 1
        return 0
    if neighbs == 3:
        return 1
    return 0
```

Listing 11-10: Changes to the checkNeighbors() code to include diagonal neighbors

First, we add four values ❶ to check the diagonal neighbors: [-1,-1] for the neighbor to the left and up, [1,1] for the neighbor to the right and down, and so on. Then we tell the program that if the cell is on ❷, check if it has two or three neighbors that are also on. If so, we tell the program to return 1, and if not, we tell the program to return 0. Otherwise, if the cell is off, we tell it to check if it has three neighbors that are on. If it does, return 1; if doesn't, return 0.

Then we place living cells randomly around the grid, so we have to import the choice() function from Python's random module. Add this line to the top of the program:

```
from random import choice
```

Then we use the choice() function to randomly choose whether a new Cell is on or off. So all we have to do is change the append line in the createCellList() function to the following:

```
newList [j].append(Cell(i,j,choice([0,1])))
```

Now we no longer need the generation code from the previous file. The remaining code in the draw() function looks like this:

```
def draw():
    global cellList
    frameRate(10)
    cellList = update(cellList)
    for row in cellList:
        for cell in row:
            cell.display()
```

Run this code, and you'll see a wild, dynamic game play out, where organisms are moving, morphing, splitting, and interacting with other organisms, like in Figure 11-10.

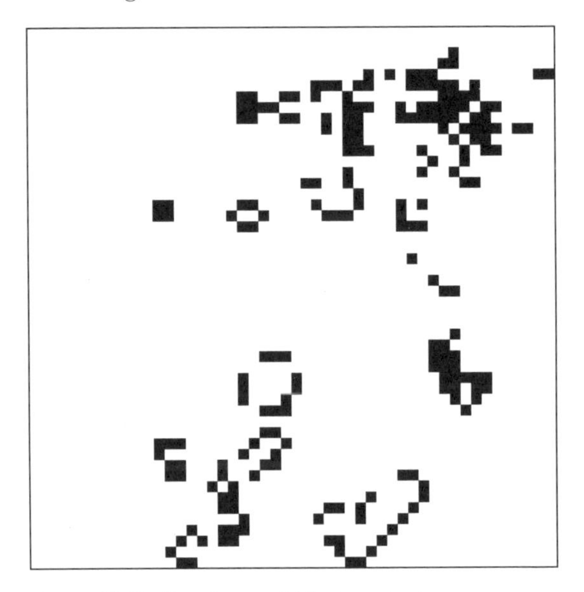

Figure 11-10: The Game of Life in action!

It's interesting how the "clouds" of cells morph, move, and collide with other clouds (families? colonies?). Some organisms wander around the screen until, eventually, the grid will settle into a kind of equilibrium. Figure 11-11 shows an example of that kind of equilibrium.

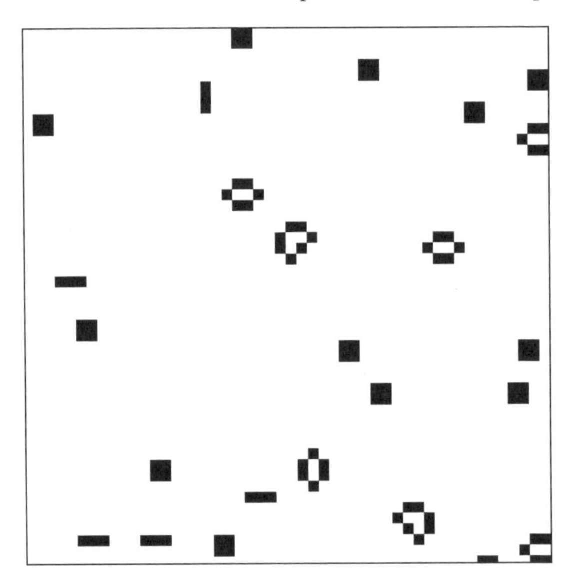

Figure 11-11: An example of the Game of Life that has entered a stable state

In this example of a state of equilibrium, some shapes appear stable and unmoving, while other shapes become stuck in repeating patterns.

THE ELEMENTARY CELLULAR AUTOMATON

This last CA is really cool and involves a little more math, but it's still a simple pattern that's extended, though only in one dimension (which is why it's called an "elementary CA"). We start off with one row of cells and set the middle cell's state to one, as shown in Figure 11-12.

Figure 11-12: The first row of an elementary CA

This is easy to code. Start a new Processing sketch and call it *elementaryCA.pyde*. The code to draw the first row of cells is shown in Listing 11-11.

elementaryCA.pyde

```
❶ #CA variables
w = 50
rows = 1
cols = 11

def setup():
    global cells
    size(600,600)
    #first row:
  ❷ cells = []
    for r in range(rows):
        cells.append([])
        for c in range(cols):
            cells[r].append(0)
  ❸ cells[0][cols//2] = 1

def draw():
    background(255) #white
    #draw the CA
    for i, cell in enumerate(cells): #rows
        for j, v in enumerate(cell): #columns
          ❹ if v == 1:
                fill(0)
            else: fill(255)
          ❺ rect(j*w-(cols*w-width)/2,w*i,w,w)
```

Listing 11-11: Drawing the first row (generation) of the elementary CA

First, we declare a few important variables ❶, such as the size of each cell and the number of rows and columns in our CA. Next, we start our cells list ❷. We create `rows` number of rows and append `cols` number of 0's in each list inside `cells`. We set the middle cell in the row to 1 (or on) ❸. In the `draw()` function, we loop through the rows (there will be more than one row soon!) and columns using `enumerate`. We check if the element is a 1, and if so, we color it black ❹. Otherwise, we color it white. Finally, we draw the square for the cell ❺. The x-value looks a bit complicated, but this just makes sure the CA is always centered.

When you run this code, you should see what's shown in Figure 11-12: a row of cells with one "on" cell in the center. The state of the cells in the next row of the CA will depend on the rules we set up for a cell and its two neighbors. How many possibilities are there? Each cell has two possible states (1 or 0, or "on" or "off") so that's two states for the left neighbor, two for the center cell, and two for the right neighbor. That's $2 \times 2 \times 2 = 8$ possibilities. All the combinations are shown in Figure 11-13.

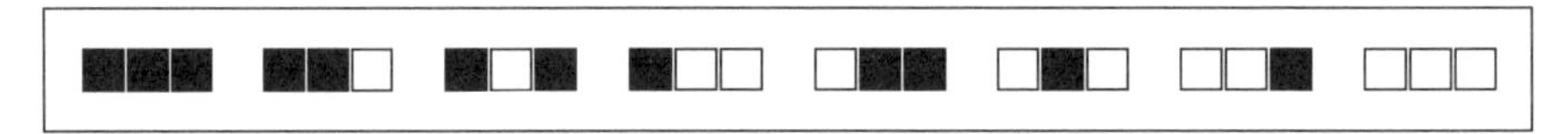

Figure 11-13: All the possible combinations of a cell and its two neighbors

The first possibility is that the center cell is on and both its neighbors are on. The next possibility is that the center cell is on, the left neighbor is on, and the right neighbor is off—and so on. This order is very important. (Do you see the pattern?) How are we going to describe these possibilities to the computer program? We could write eight conditional statements like this one:

```
if left == 1 and me == 1 and right == 1:
```

But there's an easier way. In A *New Kind of Science*, Stephen Wolfram assigns numbers to the possibilities according to the binary number the three cells represent. Keeping in mind that 1 is on and 0 is off, you can see that 111 is 7 in binary, 110 is 6 in binary, and so on, as illustrated in Figure 11-14.

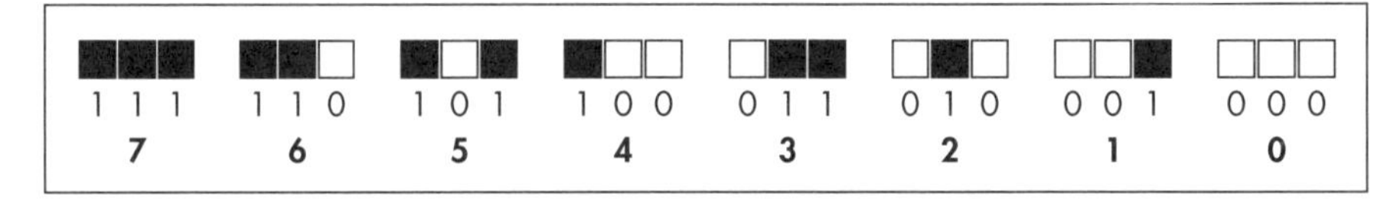

Figure 11-14: The numbering method for the eight possibilities

Now that we've numbered each possibility, we can create a rule set—that is, a list that will contain the rules for what to do with each possibility in the next generation. Notice the numbers are like the indices of a list, except backwards. We can easily get around that. We can assign a result to each one randomly or because of some plan. Figure 11-15 shows one set of results.

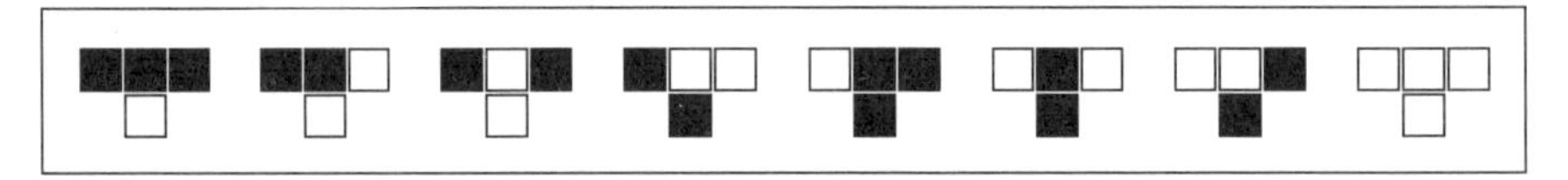

Figure 11-15: A set of results assigned to each possibility in the CA

The box under each possibility signifies the result, or the state of the cell in the next generation of the CA. The white box under "possibility 7" on the left means "if the cell is on and both of its neighbors are on, it will be off in the next generation." Same for the next two possibilities (which don't exist in our CA so far): the result is "off." As illustrated earlier in Figure 11-12, we have a lot of "off" cells surrounded by "off" cells, which is the possibility shown on the far right of Figure 11-14: three white squares. In this case, the cell in the middle will be off in the next generation. We also have one "on" cell surrounded by two "off" cells (possibility 5). In the next generation, the cell will be on. We'll use 0's and 1's for our `ruleset` list, as illustrated in Figure 11-16.

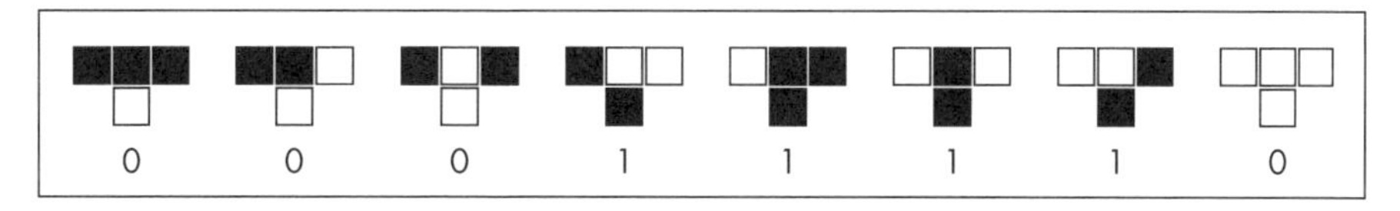

Figure 11-16: Putting the rules for generating the next row into a list

We'll collect all these numbers into a list called `ruleset`, which we'll add just before the `setup()` function:

```
ruleset = [0,0,0,1,1,1,1,0]
```

The order of the possibilities is important because this rule set is referred to as "Rule 30" (00011110 is 30 in binary). Our task is to create the next row according to the rules. Let's create a `generate()` function that looks at the first row and generates the second row, then looks at the second row and generates the third row, and so on. Add the code shown in Listing 11-12.

elementaryCA.pyde

```
#CA variables
w = 50
❶ rows = 10
cols = 100
--snip--
ruleset = [0,0,0,1,1,1,1,0] #rule 30

❷ def rules(a,b,c):
    return ruleset[7 - (4*a + 2*b + c)]

def generate():
    for i, row in enumerate(cells): #look at first row
        for j in range(1,len(row)-1):
            left = row[j-1]
            me = row[j]
            right = row[j+1]
            if i < len(cells) - 1:
                cells[i+1][j] = rules(left,me,right)
    return cells
```

Listing 11-12: Writing the `generate()` function to generate new rows in the CA

First, we make the CA larger by updating the number of rows and columns ❶. Next, we create the `rules()` function ❷, which takes three parameters: the left neighbor's number, the current cell's number, and the right neighbor's number. The function checks the `ruleset` and returns the value for the cell in the next generation. We make use of the binary numbers, and the line `4*a + 2*b + c` converts "1,1,1" to 7 and "1,1,0" to 6, and so on. However, as you'll recall from Figure 11-15, the indices are in reverse order, so we subtract the total from 7 to get the proper index of the `ruleset`.

Add the following line to the end of the `setup()` function:

```
cells = generate()
```

This creates the full CA and not just the first row. When you run this code, you should see the first 10 rows of a CA made using "Rule 30," as illustrated in Figure 11-17.

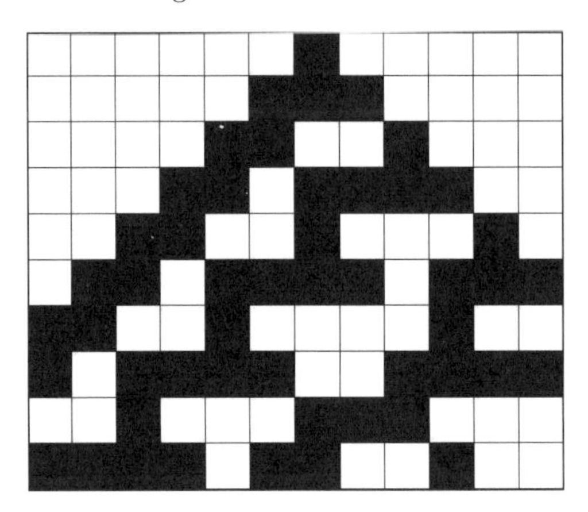

Figure 11-17: The first 10 rows of Rule 30

The program is going through each row, starting at the top, and generating the next row according to the rules we gave it in the `ruleset`. What if we keep going? Change the number of rows and columns to 1000 and the width (`w`) of each cell to 3. Add `noStroke()` to the `setup()` function to get rid of the outline of the cells, and then run the sketch. You should see what's in Figure 11-18.

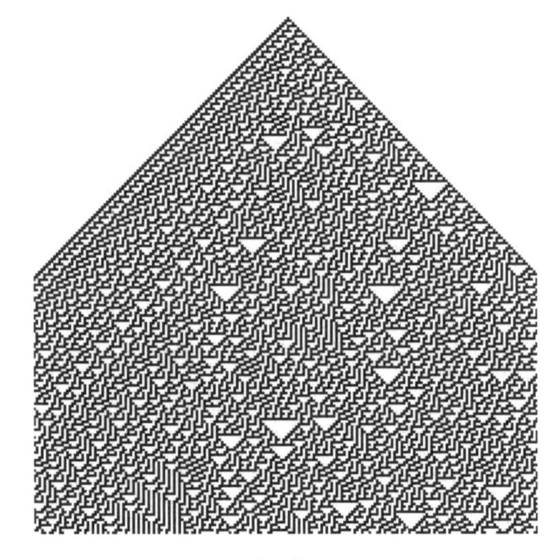

Figure 11-18: More of Rule 30

Rule 30 is a fascinating design because it's not completely random, but it's not completely regular either.

Rule 73 is also cool; in fact a woman named Fabienne Serriere programs the rule into a knitting machine to create scarves with the pattern, like in Figure 11-19. You can order scarves with this and other algorithmically generated rules on them from *https://knityak.com/*.

Figure 11-19: A scarf whose design is a cellular automaton: Rule 73!

> **EXERCISE 11-2: CHANGING THE RULE SET**
>
> Change `ruleset` to the binary form of the number 90. What does the resulting CA look like? Hint: it's a fractal.

> **EXERCISE 11-3: ZOOMING IN AND OUT**
>
> Use the `keyPressed()` function you learned about in Chapter 10 to change the value of the width variable `w` using the up and down arrow keys. This should let you zoom in and out of the CA!

SUMMARY

In this chapter, you learned to use Python to create cellular automata, or cells that act independently, according to specific rules. We wrote programs to make a huge grid of these cells follow certain rules and update themselves, generation after generation, and we created unexpectedly beautiful designs and surprisingly life-like behavior.

In the next chapter, we'll create virtual organisms that solve problems for us! These organisms will be able to guess a secret phrase and find the shortest route through a bunch of cities just by evolving better and better solutions.

12

SOLVING PROBLEMS USING GENETIC ALGORITHMS

Steve: We're lost.
Mike: How lost are we?

When many people think of math, they think of equations and operations that are "set in stone" and answers that are either right or wrong. They might be surprised to learn how much guessing and checking we've done in our algebra explorations already.

In this chapter, you learn to crack passwords and hidden messages in an indirect fashion. It's kind of like the "guess-and-check" method of Chapter 4, where we just plugged a bunch of integers into an equation and if any made the equation true, we printed them out. This time, we'll guess a bunch of values, not just one. It's not the most elegant way of solving a problem, but with a computer at our disposal, sometimes brute force works best.

To figure out our secret phrase, we generate guesses and then rate them on how well they match the target. But here's where we depart from a guess-and-check method: we keep the best guesses and mutate them, randomly, again and again until we uncover the message. The program won't know which letters are right and which letters are wrong, but we get closer and closer by mutating the best guess we've made so far. Although this method might not seem promising right now, you'll see that it helps crack the code surprisingly quickly. This method is called a *genetic algorithm*, which computer scientists use to find solutions to problems based on the theory of natural selection and evolutionary biology. It was inspired by biological organisms that adapt and mutate, and the way they build on tiny advantages, as we saw in the Sheep model in Chapter 9, on Classes.

For more complicated problems, however, random mutating won't be enough to solve our problem. In those cases, we add *crossover*, which we use to combine the most fit organisms (or best guesses) to improve their likelihood of cracking the code, just like how the fittest organisms are more likely to pass down a combination of their genetic material. All this activity, other than the scoring, will be fairly random, so it might be surprising that our genetic algorithms work so well.

USING A GENETIC ALGORITHM TO GUESS PHRASES

Open IDLE and create a new file called *geneticQuote.py*. Instead of guessing a number like in Chapter 4, this program tries to guess a secret phrase. All we have to tell the program is the number of characters it guessed correctly—not where or which characters, just how many.

Our program is going to be able to do much better than guess short passwords.

WRITING THE MAKELIST() FUNCTION

To see how this works, let's create a target phrase. Here's a long sentence that my son came up with from the comic book *Naruto*:

```
target = "I never go back on my word, because that is my Ninja way."
```

In English, we have a bunch of characters we can choose from: lowercase letters, uppercase letters, a space, and some punctuation.

```
characters = " abcdefghijklmnopqrstuvwxyzABCDEFGHIJKLMNOPQRSTUVWXYZ.',?!"
```

Let's create a function called `makeList()` that will randomly create a list of characters that's the same length as `target`. Later, when we try to guess what the target phrase is, we'll score the guess by comparing it character by character with the target. A higher score means a guess is closer to the target. Then, we'll randomly change one of the characters in that guess to see if that increases its score. It seems surprising that such a random method will ever get us to the exact target phrase, but it will.

First, import the `random` module and write the `makeList()` function, as shown in Listing 12-1.

geneticQuote.py

```
import random

target = "I never go back on my word, because that is my Ninja way."
characters = " abcdefghijklmnopqrstuvwxyzABCDEFGHIJKLMNOPQRSTUVWXYZ.',?!"

def makeList():
    '''Returns a list of characters the same length
    as the target'''
    charList = [] #empty list to fill with random characters
    for i in range(len(target)):
        charList.append(random.choice(characters))
    return charList
```

Listing 12-1: Writing the `makeList()` function to create a list of random characters that's the same length as the target

Here, we create an empty list called `charList` and loop over the list the same number of times as there are characters in the target. On each loop,the program puts a random character from `characters` into `charList`. Once the loop is done, it returns `charList`. Let's test it to make sure it works.

TESTING THE MAKELIST() FUNCTION

First, let's find out what the length of the target is, and check that our random list is the same length:

```
>>> len(target)
57
>>> newList = makeList()
>>> newList
['p', 'H', 'Z', '!', 'R', 'i', 'e', 'j', 'c', 'F', 'a', 'u', 'F', 'y', '.',
'w', 'u', '.', 'H', 'W', 'w', 'P', 'Z', 'D', 'D', 'E', 'H', 'N', 'f', ' ',
'W', 'S', 'A', 'B', ',', 'w', '?', 'K', 'b', 'N', 'f', 'k', 'g', 'Q', 'T',
'n', 'Q', 'H', 'o', 'r', 'G', 'h', 'w', 'l', 'l', 'W', 'd']
>>> len(newList)
57
```

We measured the length of the `target` list, and it's 57 characters long. Our new list is the same length, 57 characters. Why make a list instead of a string? We make a list because lists are sometimes easier to work with than strings. For example, you can't simply replace a character in a string with another character. But in a list you can, as you can see here:

```
>>> a = "Hello"
>>> a[0] = "J"
Traceback (most recent call last):
  File "<pyshell#16>", line 1, in <module>
    a[0] = "J"
TypeError: 'str' object does not support item assignment
>>> b = ["H","e","l","l","o"]
>>> b[0] = "J"
>>> b
['J', 'e', 'l', 'l', 'o']
```

In this example, when we try to replace the first item in the "Hello" string with "J", Python doesn't let us, and we get an error. Doing the same thing using a list, however, is no problem.

In the case of our *geneticQuote.py* program, we want to see the random quote as a string because that's easier to read. Here's how to print out a list as a string, using Python's join() function:

```
>>> print(''.join(newList))
pHZ!RiejcFauFy.wu.HWwPZDDEHNf WSAB,w?KbNfkgQTnQHorGhwllWd
```

Those are all the characters in newList, but in string form. It doesn't look like a very promising start!

WRITING THE SCORE() FUNCTION

Now let's write a function called score() to score each guess by comparing it character by character with the target, like in Listing 12-2.

geneticQuote.py

```
def score(mylist):
    '''Returns one integer: the number of matches with target'''
    matches = 0
    for i in range(len(target)):
        if mylist[i] == target[i]:
            matches += 1
    return matches
```

Listing 12-2: Writing the score() function for scoring a guess

The score() function takes each item in a list we feed it (mylist) and checks if the first character of mylist matches the first character of the target list. Then the function checks whether the second characters match, and so on. For each character matched, we increment matches by 1. In the end, this function returns a single number, not which ones are right, so we don't actually know *which* characters we got right!

What's our score?

```
>>> newList = makeList()
>>> score(newList)
0
```

Our first guess was a total strikeout. Not a single match!

WRITING THE MUTATE() FUNCTION

Now we'll write a function to mutate a list by randomly changing one character. This will allow our program to "make guesses" until we get closer to the target phrase we're trying to guess. The code is in Listing 12-3.

geneticQuote.py

```
def mutate(mylist):
    '''Returns mylist with one letter changed'''
    newlist = list(mylist)
```

```
    new_letter = random.choice(characters)
    index = random.randint(0,len(target)-1)
    newlist[index] = new_letter
    return newlist
```

Listing 12-3: Writing the `mutate()` function for changing one character in a list

First, we copy the elements of the list to a variable called `newlist`. We then randomly choose a character from the `characters` list to be the new letter that will replace one of the existing characters. We randomly choose a number between 0 and the length of the target to be the index of the letter we replace. Then we set the character in `newlist` at that index to be the new letter. This process repeats over and over again in a loop. If the new list has a higher score, it'll become the "best" list, and the best list will keep getting mutated in the hope of improving its score even more.

GENERATING A RANDOM NUMBER

Starting off the program after all the function definitions, we make sure of our randomness by calling `random.seed()`. Calling `random.seed()` resets the random number generator to the present time. Then we make a list of characters and, since the first list is the best one so far, declare it the best list. Its score will be the best score.

geneticQuote.py

```
random.seed()
bestList = makeList()
bestScore = score(bestList)
```

We keep track of how many guesses we've made:

```
guesses = 0
```

Now we start an infinite loop that will mutate `bestList` to make a new guess. We calculate its score and increment the `guesses` variable:

```
while True:
    guess = mutate(bestList)
    guessScore = score(guess)
    guesses += 1
```

If the score of the new guess is less than or equal to the best score so far, the program can "continue," as shown next. That means it will go back to the beginning of the loop, since it wasn't a good guess, and we don't need to do anything else with it.

```
    if guessScore <= bestScore:
        continue
```

If we're still in the loop, that means the guess is good enough to print out. We print its score, too. We can print the list (as a string), the score, and how many total guesses were made. If the score of the new guess is the same as the length of the target, then we've solved the quote and we can break out of the loop:

```
    print(''.join(guess),guessScore,guesses)
    if guessScore == len(target):
        break
```

Otherwise, the new guess must be better than the best list so far, but not perfect yet, so we can declare it the best list and save its score as the best score:

```
    bestList = list(guess)
    bestScore = guessScore
```

Listing 12-4 shows the entire code for the *geneticQuote.py* program.

geneticQuote.py

```
import random

target = "I never go back on my word, because that is my Ninja way."
characters = " abcdefghijklmnopqrstuvwxyzABCDEFGHIJKLMNOPQRSTUVWXYZ.',?!"

#function to create a "guess" list of characters the same length as target
def makeList():
    '''Returns a list of characters the same length
    as the target'''
    charList = [] #empty list to fill with random characters
    for i in range(len(target)):
        charList.append(random.choice(characters))
    return charList

#function to "score" the guess list by comparing it to target
def score(mylist):
    '''Returns one integer: the number of matches with target'''
    matches = 0
    for i in range(len(target)):
        if mylist[i] == target[i]:
            matches += 1
    return matches

#function to "mutate" a list by randomly changing one letter
def mutate(mylist):
    '''Returns mylist with one letter changed'''
    newlist = list(mylist)
    new_letter = random.choice(characters)
    index = random.randint(0,len(target)-1)
    newlist[index] = new_letter
    return newlist

#create a list, set the list to be the bestList
#set the score of bestList to be the bestScore
```

```
random.seed()
bestList = makeList()
bestScore = score(bestList)

guesses = 0

#make an infinite loop that will create a mutation
#of the bestList, score it
while True:
    guess = mutate(bestList)
    guessScore = score(guess)
    guesses += 1

#if the score of the newList is lower than the bestList,
#"continue" on to the next iteration of the loop
    if guessScore <= bestScore:
        continue

#if the score of the newlist is the optimal score,
#print the list and break out of the loop
    print(''.join(guess),guessScore,guesses)
    if guessScore == len(target):
        break

#otherwise, set the bestList to the value of the newList
#and the bestScore to be the value of the score of the newList
    bestList = list(guess)
    bestScore = guessScore
```

Listing 12-4: The complete code for the geneticQuote.py *program*

Now when we run this, we get a very fast solution, with all the guesses that improved the score printed out.

```
i.fpzgPG.'kHT!NW WXxM?rCcdsRCiRGe.LWVZzhJe zSzuWKV.FfaCAV 1 178
i.fpzgPG.'kHT!N  WXxM?rCcdsRCiRGe.LWVZzhJe zSzuWKV.FfaCAV 2 237
i.fpzgPG.'kHT!N  WXxM?rCcdsRCiRGe.LWVZzhJe zSzuWKV.FfwCAV 3 266
i fpzgPG.'kHT!N  WXxM?rCcdsRCiRGe.LWVZzhJe zSzuWKV.FfwCAV 4 324
--snip--
I nevgP go back on my word, because that is my Ninja way. 55 8936
I neveP go back on my word, because that is my Ninja way. 56 10019
I never go back on my word, because that is my Ninja way. 57 16028
```

This output shows that the final score was 57, and it took 16,028 total guesses to match the quote exactly. Notice on the first line of output that 178 guesses were needed to get a score of 1! There are more efficient ways of guessing a quote, but I wanted to introduce the idea of genetic algorithms using an easy example. The point was to show how a method of scoring guesses and randomly mutating the "best guess so far" could produce accurate results in a surprisingly short amount of time.

Now, you can use this idea of scoring and mutating thousands of random guesses to solve other problems, too.

SOLVING THE TRAVELING SALESPERSON PROBLEM (TSP)

One of my students was unimpressed with the quote-guessing program because "we already know what the quote is." So let's use a genetic algorithm to solve a problem we don't already know the solution for. *The Traveling Salesperson Problem*, or *TSP* for short, is an age-old brainteaser that is easy to understand but can become very difficult to solve. A salesperson has to travel to a given number of cities, and the goal is to find the route with the shortest distance. Sounds easy? And with a computer, we should simply be able to run all the possible routes through a program and measure their distances, right?

It turns out, above a certain number of cities, the computational complexity gets too much even for today's supercomputers. Let's see how many possible routes there are when you have six cities, as shown in Figure 12-1.

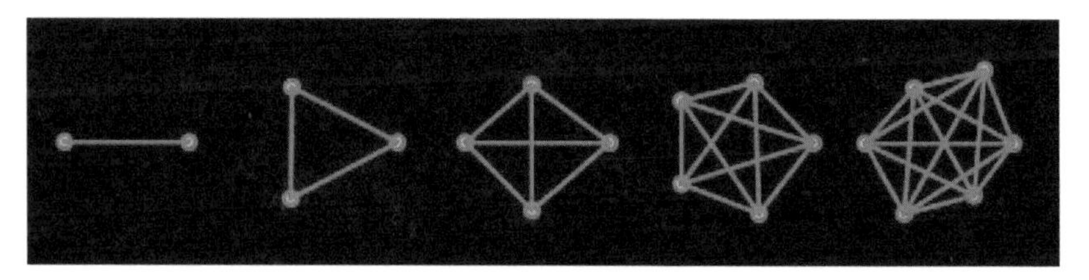

Figure 12-1: The number of paths between n *cities for* n *between 2 and 6*

When there are two or three cities, there's only one possible route. Add a fourth city, and it could be visited between any of the previous three, so multiply the previous number of routes by 3. So between four cities there are three possible routes. Add a fifth city, and it could be visited between any of the previous four, so there are four times as many as the previous step, so 12 possible routes. See the pattern? Between *n* cities, there are

$$\frac{(n-1)!}{2}$$

possible routes. So between 10 cities there are 181,440 possible routes. Between 20 cities, there are 60,822,550,204,416,000 routes. What's after a trillion? Even if a computer can check a million routes per second, it would still take almost 2,000 years to calculate. That's too slow for our purposes. There must be a better way.

USING GENETIC ALGORITHMS

Similar to our quote-guessing program, we're going to create an object with a route in its "genes" and then score its route by how short it is. The best route will then be mutated randomly, and we'll score its mutation. We could take a bunch of "best routes," splice together their lists, and score their "offspring." The best part of this exploration is we *don't* know the answer already. We could give the program a set of cities and their locations, or just have it randomly draw cities and try to optimize the route.

Open a new Processing sketch and call it *travelingSalesperson.pyde.* The first thing we should create is a `City` object. Each city will have its own x- and y-coordinate and a number we use to identify it. That way, we can define a route using a list of city numbers. For example, [5,3,0,2,4,1] means you start at city 5 and go to city 3, then city 0, and so on. The rules are the salesperson has to finally return to the first city. Listing 12-5 shows the `City` class.

travelingSalesperson.pyde

```
class City:
    def __init__(self,x,y,num):
        self.x = x
        self.y = y
        self.number = num #identifying number

    def display(self):
        fill(0,255,255) #sky blue
        ellipse(self.x,self.y,10,10)
        noFill()
```

Listing 12-5: Writing the `City` class for the travelingSalesperson.pyde *program*

When initializing `City`, we get an x- and y-coordinate and give each `City` its own (`self`) x- and y-component. We also get a number that's the city's identifying number. In the `display()` method, we choose a color (sky blue, in this case) and create an ellipse at that location. We turn off the fill after drawing the city with the `noFill()` function, since no other shapes need to be filled in with color.

Let's make sure that works. Let's create the `setup()` function, declaring a size for the display window and creating an instance of our `City` class. Remember, we have to give it a location of two coordinates and an identifying number as in Listing 12-6.

```
def setup():
    size(600,600)
    background(0)
    city0 = City(100,200,0)
    city0.display()
```

Listing 12-6: Writing the `setup()` function for creating one city

Run this, and you'll see your first city (see Figure 12-2)!

Figure 12-2: The first city

It might help to have the city display its number above it. To do that, add this to the city's `display()` method, just before `noFill()`:

```
        textSize(20)
        text(self.number,self.x-10,self.y-10)
```

We declare the size of the text using Processing's built-in `textSize()` function. Then we use the `text()` function to tell the program what to print (the number of the city) and where to print it (10 pixels to the left and above the city). While we're creating cities, let's start a `cities` list and put a few more cities on the screen in random locations. To use methods from the `random` module, we have to import `random` at the top of the file:

```
import random
```

Now we can update our `setup()` function like in Listing 12-7.

travelingSalesperson.pyde

```
cities = []

def setup():
    size(600,600)
    background(0)
    for i in range(6):
        cities.append(City(random.randint(50,width-50),
                           random.randint(50,height-50),i))

    for city in cities:
        city.display()
```

Listing 12-7: Writing the setup() function for creating six random cities

In the `setup()` function, we've added a loop to run six times. It adds a `City` object at a random location on the screen 50 units from the edges. The next loop iterates over all the elements in the `cities` list and displays each one. Run this, and you'll see six cities in random locations, labeled with their ID numbers, as in Figure 12-3.

Figure 12-3: Six cities, labeled with their numbers

Now let's think about the route between the cities. We put the `City` objects (containing their locations and numbers) into the `cities` list, and eventually that list of numbers (our "genetic material") will consist of the city numbers in a certain order. So the `Route` object needs a random list of numbers, too: a random sequence of all the city numbers. Of course, the numbers will be range from 0 to 1 less than the number of cities. We don't want to keep changing numbers here and there in our code whenever we want to change the number of cities, so we'll create a variable for the number of cities. Put this line at the beginning of the file, before the `City` class:

```
N_CITIES = 10
```

Why is `N_CITIES` in all capital letters? Throughout all the code, we won't be changing the number of cities. So it's not really a variable; instead, it's a constant. It's customary in Python to capitalize constant names to set them apart from variables. This doesn't change the way Python deals with them at all; variables with capitalized names can still be changed. So be careful.

We'll use `N_CITIES` wherever we would be using the total number of cities, and we'll only need to change the value once! Place the code shown in Listing 12-8 after the `City` class.

```
class Route:
    def __init__(self):
        self.distance = 0
        #put cities in a list in order:
        self.cityNums = random.sample(list(range(N_CITIES)),N_CITIES)
```

Listing 12-8: The `Route` *class*

First, we set the route's distance (or length, but `length` is a keyword in Processing) to zero, and then we create a `cityNums` list that puts the numbers of the cities in a random order for that route.

You can use the `random` module's `sample()` function to give Python a list and then sample a number of items from that list by telling it how many items to choose randomly. It's like `choice()`, but it won't select an item more than once. In probability, it's called "sampling without replacement." Enter the following in IDLE to see how sampling works:

```
>>> n = list(range(10))
>>> n
[0, 1, 2, 3, 4, 5, 6, 7, 8, 9]
>>> import random
>>> x = random.sample(n,5)
>>> x
[2, 0, 5, 3, 8]
```

Here, we create a list called `n` of the numbers between 0 and 9 by calling `range(10)` and converting it (it's a "generator") into a list. We then import the `random` module and ask Python to use the `sample()` function to pick a sample of five items from list `n` and save them to list `x`. In our `Route` code in Listing 12-8, since the variable `N_CITIES`, representing the number of cities, is 10, we're choosing 10 numbers at random using `range(10)`, the numbers 0 to 9, and assigning them to the `Route`'s `cityNums` property.

And how will this display? Let's draw purple lines between the cities. You can use any color you'd prefer.

Drawing lines between cities like this should remind you of drawing lines between the points on a graph in algebra or trigonometry lessons. The only difference is now at the end of the graph we have to return to the starting point. Remember using `beginShape`, `vertex`, and `endShape` in Chapter 6? Just like we used lines to draw a shape, we'll draw the `Route` object as the outline of a shape, except this time we just won't fill it in. Using `endshape(CLOSE)` will automatically close the loop! Add the code in Listing 12-9 to the `Route` class.

```
    def display(self):
        strokeWeight(3)
        stroke(255,0,255) #purple
        beginShape()
        for i in self.cityNums:
            vertex(cities[i].x,cities[i].y)
            #then display the cities and their numbers
```

```
        cities[i].display()
    endShape(CLOSE)
```

Listing 12-9: Writing the `display` method of the `Route` class

The loop makes every city in the `Route`'s `cityNums` list a vertex of a polygon. The route is the outline of the polygon. Notice that inside the `Route`'s `display()` method we call the city's `display()` method. That way, we don't have to manually command the cities to display separately.

In the `setup()` function, we'll create a `Route` object with the `cities` list and a list of numbers as arguments. Then we'll display it. The last two lines of code at the bottom of Listing 12-10 do this.

```
def setup():
    size(600,600)
    background(0)
    for i in range(N_CITIES):
        cities.append(City(random.randint(50,width-50),
                           random.randint(50,height-50),i))
    route1 = Route()
    route1.display()
```

Listing 12-10: Displaying a route

Run this, and you'll see a path between the cities, in random order, as shown in Figure 12-4.

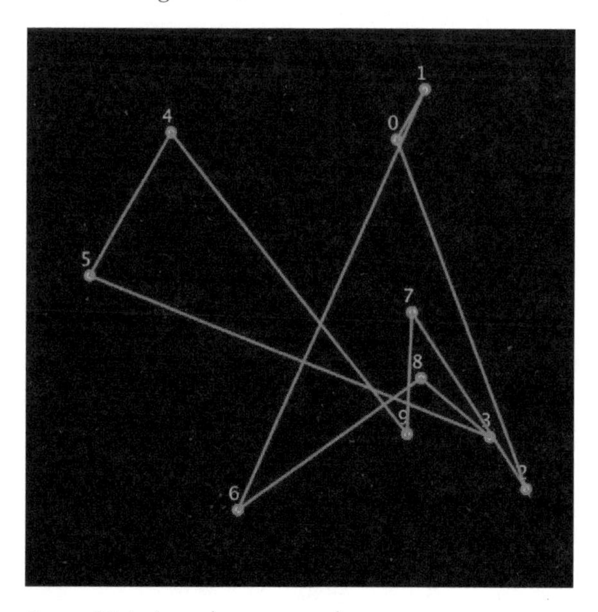

Figure 12-4: A random route order

To change the number of cities, simply change the first line, where we declare `N_CITIES`, to a different number and then run the program. Figure 12-5 shows my output for `N_CITIES = 7`.

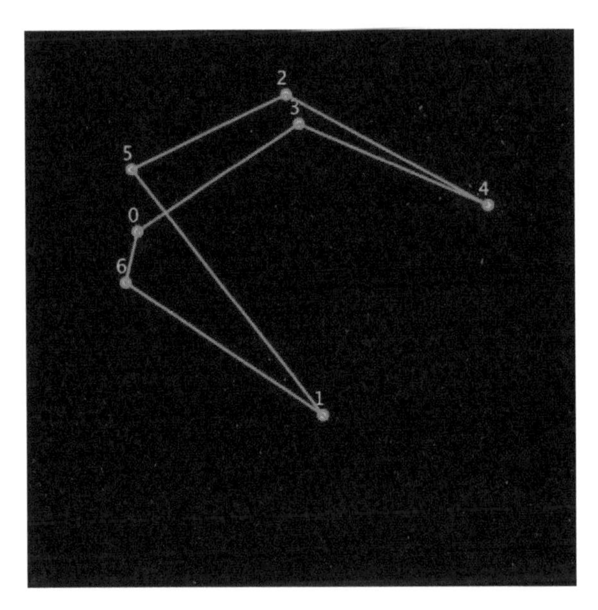

Figure 12-5: A route with seven cities

Now that you can create and display routes, let's write a function to measure the distance of each route.

WRITING THE CALCLENGTH() METHOD

The `Route` object has a `distance` property that's set to zero when it's created. Each `Route` object also has a list of cities, in order, called `cityNums`. We just have to loop through the `cityNums` list and keep a running total of the distances between each pair of cities. No problem for cities 0 to 4, but we also need to calculate the distance from the last city back to the first one.

Listing 12-11 shows the code for the `calcLength()` method, which goes inside the `Route` object.

```
def calcLength(self):
    self.distance = 0
    for i,num in enumerate(self.cityNums):
    # find the distance from the current city to the previous city
        self.distance += dist(cities[num].x,
                              cities[num].y,
                              cities[self.cityNums[i-1]].x,
                              cities[self.cityNums[i-1]].y)
    return self.distance
```

Listing 12-11: Calculating a `Route`*'s length*

First, we zero out the `distance` property of the `Route` so every time we call this method it'll start at zero. We use the `enumerate()` function so we can get not just the number in the `cityNums` list but also its index. We then increment the `distance` property by the distance from the current city (`num`) to the previous city (`self.cityNums[i-1]`). Next, let's add this line of code to the end of our `setup()` function:

```
    println(route1.calcLength())
```

We can now see the total distance covered by the salesperson in the console, like in Figure 12-6.

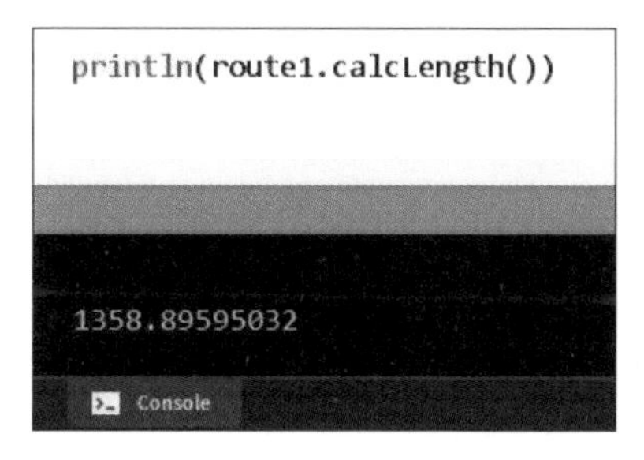

Figure 12-6: We've calculated the distance . . . I think.

Is this really the distance? Let's make sure.

TESTING THE CALCLENGTH() METHOD

Let's give the program an easy route that's a square of sidelength 200 and check the distance. First, we change our constant for the number of cities to 4:

```
N_CITIES = 4
```

Next, we change the `setup()` function to what's shown in Listing 12-12.

```
cities = [City(100,100,0), City(300,100,1),
                   City(300,300,2), City(100,300,3)]

def setup():
    size(600,600)
    background(0)
    '''for i in range(N_CITIES):
        cities.append(City(random.randint(0,width),
                           random.randint(0,height),i))'''
    route1 = Route()
    route1.cityNums = [0,1,2,3]
    route1.display()
    println(route1.calcLength())
```

Listing 12-12: Creating a `Route` manually to test the `calcLength()` method

We comment out the loop to create cities at random, because we'll go back to it after checking the `calcLength()` method. We create a new `cities` list containing the vertices of a square of sidelength 200. We also declare the `cityNums` list for `route1`; otherwise, it would randomly mix the cities. We expect the length of this `Route` to be 800.

When we run the code, we see what's in Figure 12-7.

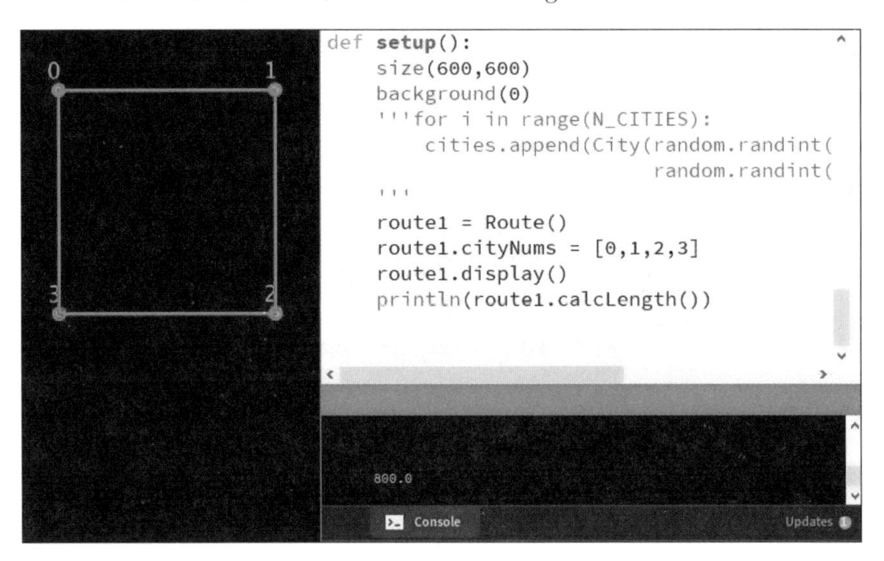

Figure 12-7: The `calcLength()` method works!

It's 800 units, as predicted! You can try some rectangles or some other easy-to-verify routes.

RANDOM ROUTES

In order to find the shortest possible route to a destination, we need to find all the possible routes. To do this, we need our infinite loop and Processing's built-in `draw()` function. We'll move the route code from `setup()` to the `draw()` function. We'll also create a bunch of random routes and display them and their length. The entire code is shown in Listing 12-13.

travelingSalesperson.pyde

```
import random

N_CITIES = 10

class City:
    def __init__(self,x,y,num):
        self.x = x
        self.y = y
        self.number = num #identifying number

    def display(self):
        fill(0,255,255) #sky blue
```

```
        ellipse(self.x,self.y,10,10)
        textSize(20)
        text(self.number,self.x-10,self.y-10)
        noFill()

class Route:
    def __init__(self):
        self.distance = 0
        #put cities in a list in numList order:
        self.cityNums = random.sample(list(range(N_CITIES)),N_CITIES)

    def display(self):
        strokeWeight(3)
        stroke(255,0,255) #purple
        beginShape()
        for i in self.cityNums:
            vertex(cities[i].x,cities[i].y)
            #then display the cities and their numbers
            cities[i].display()
        endShape(CLOSE)

    def calcLength(self):
        self.distance = 0
        for i,num in enumerate(self.cityNums):
        # find the distance to the previous city
            self.distance += dist(cities[num].x,
                                  cities[num].y,
                                  cities[self.cityNums[i-1]].x,
                                  cities[self.cityNums[i-1]].y)
        return self.distance

cities = []

def setup():
    size(600,600)
    for i in range(N_CITIES):
        cities.append(City(random.randint(50,width-50),
                           random.randint(50,height-50),i))

def draw():
    background(0)
    route1 = Route()
    route1.display()
    println(route1.calcLength())
```

Listing 12-13: Creating and displaying random routes

When you run this, you should see a bunch of routes being displayed and a bunch of numbers being printed to the console.

But we're really only interested in keeping the best (shortest) route, so we'll add some code to save the "bestRoute" and check the new random routes. Change setup() and draw() to what's shown in Listing 12-14.

```
cities = []
random_improvements = 0
mutated_improvements = 0

def setup():
    global best, record_distance
    size(600,600)
    for i in range(N_CITIES):
        cities.append(City(random.randint(50,width-50),
                           random.randint(50,height-50),i))
    best = Route()
    record_distance = best.calcLength()

def draw():
    global best, record_distance, random_improvements
    background(0)
    best.display()
    println(record_distance)
    println("random: "+str(random_improvements))
    route1 = Route()
    length1 = route1.calcLength()
    if length1 < record_distance:
        record_distance = length1
        best = route1

        random_improvements += 1
```

Listing 12-14: Keeping track of random improvements

Before the `setup()` function, we create a variable to count the number of random improvements that are made by the program. At the same time, we create a variable we'll use in a few steps to count the mutated improvements.

In `setup()`, we created `route1` to be the first `Route`, we named it the "best route," and we named its distance the `record_distance`. Since we want to share these variables with other functions, we declare them to be global variables at the beginning of the function.

In `draw()`, we keep generating new random routes and checking if they're better than the one we think is the best route so far. Since we're using only 10 cities, this could pay off with an optimal solution, if we leave it running a while. You'll see that it only requires around a dozen random improvements. But, remember, there are only 181,440 unique routes through 10 cities. One 10-city route is shown in Figure 12-8.

If you change the number of cities to 20, however, your program will just keep running, for days if you let it, and will probably not get close to an optimal solution. We need to start using the idea from the phrase-guessing program at the beginning of the chapter of scoring our guesses and mutating the best ones. Unlike before, we'll create a "mating pool" of the best routes and combine their number lists as if they were genes.

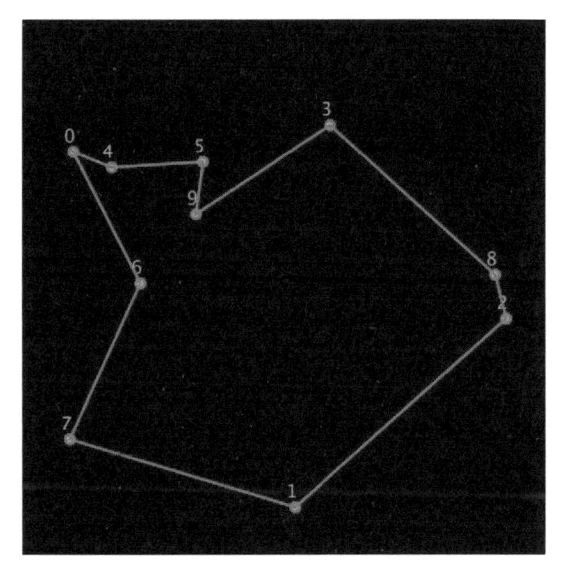

Figure 12-8: Finding an optimal route randomly—if you can wait a few minutes

APPLYING THE PHRASE-GUESSING MUTATION IDEA

The list of numbers (the cities the salesperson will visit in order) will be the genetic material of the `Route`. First, we see how well some randomly mutated routes solve the Traveling Salesman Problem (just like with our phrase-guessing programs) and then we mutate and "mate" the better routes with each other to (hopefully) create a more optimal route.

MUTATING TWO NUMBERS IN A LIST

Let's write a method to randomly mutate two of the numbers in a `Route` object's `cityNums` list. It's really just a swap. You can probably guess how we'll randomly choose two numbers and make the city numbers that have those indices in the list trade places.

Python has a unique notation for swapping the values of two numbers. You can swap two numbers without creating a temporary variable. For example, if you enter the code in Listing 12-15 in IDLE, it wouldn't work.

```
>>> x = 2
>>> y = 3
>>> x = y
>>> y = x
>>> x
3
>>> y
3
```

Listing 12-15: The wrong way to swap the values of variables

When you change the value of x to be the same as y by entering x = y, they both become 3. Now when you try to set y to be the same as x, it's not set to the original value of x (2), but the current value of x, which is 3. So both variables ended up as 3.

But you *can* swap the values on the same line, like this:

```
>>> x = 2
>>> y = 3
>>> x,y = y,x
>>> x
3
>>> y
2
```

Swapping the values of two variables like this is very useful for the mutating we're about to do. Instead of limiting the swapping to only two numbers, we can mutate more cities. We can put the swapping in a loop so the program will choose any number of cities and swap the first two numbers, then the next pair, and so on. The code for the `mutateN()` method is shown in Listing 12-16.

```
def mutateN(self,num):
    indices = random.sample(list(range(N_CITIES)),num)
    child = Route()
    child.cityNums = self.cityNums[::]
    for i in range(num-1):
        child.cityNums[indices[i]],child.cityNums[indices[(i+1)%num]] = \
        child.cityNums[indices[(i+1)%num]],child.cityNums[indices[i]]
    return child
```

Listing 12-16: Writing the `mutateN()` method, for mutating any number of cities

We give the `mutateN()` method `num`, a number of cities to swap. Then the method makes a list of indices to swap by taking a random sample from the range of city numbers. It creates a "child" `Route` and copies its own city number list to the child. Then it swaps `num-1` times. If it swapped the full `num` times, the first city swapped would simply get swapped with all the other indices and end up where it started.

That long line of code is simply the `a,b = b,a` syntax we saw before, only with the two `cityNums` being swapped. The mod (`%`) operator makes sure your indices don't exceed `num`, the number of cities in your sample. So if you're swapping four cities, for example, when `i` is `4`, it changes `i + 1` from `5` to `5 % 4`, which is 1.

Next, we add a section to the end of the `draw()` function to mutate the best `Route`'s list of numbers and test the mutated `Route`'s length, as shown in Listing 12-17.

```
def draw():
    global best,record_distance,random_improvements
    global mutated_improvements
    background(0)
```

```
best.display()
println(record_distance)
println("random: "+str(random_improvements))
println("mutated: "+str(mutated_improvements))
route1 = Route()
length1 = route1.calcLength()
if length1 < record_distance:
    record_distance = length1
    best = route1
    random_improvements += 1

for i in range(2,6):
    #create a new Route
    mutated = Route()
    #set its number list to the best one
    mutated.cityNums = best.cityNums[::]
    mutated = mutated.mutateN(i) #mutate it
    length2 = mutated.calcLength()
    if length2 < record_distance:
        record_distance = length2
        best = mutated
        mutated_improvements += 1
```

Listing 12-17: Mutating the best "organism"

In the `for i in range(2,6):` loop, we're telling the program to mutate 2, 3, 4, and 5 numbers in the `number` list and check the results. Now the program often does pretty well on a 20-city route in a few seconds, like in Figure 12-9.

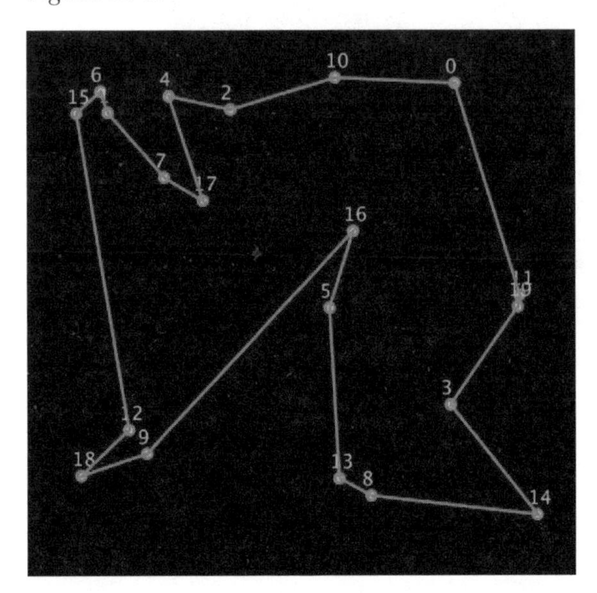

Figure 12-9: A 20-city route

The mutated "organisms" are improving the distance much better than the random ones! Figure 12-10 shows the printout.

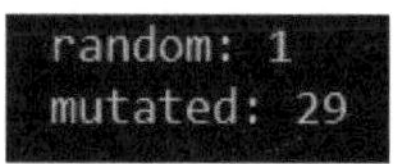

Figure 12-10: The mutations are doing much better than the random improvements!

Figure 12-10 categorizes all the improvements, and here 29 of them were due to mutations and only one was due to a randomly generated `Route`. This shows that mutating lists is better at finding the optimal route than creating new random ones. I stepped up the mutating to swap anywhere from 2 to 10 cities by changing this line:

```
for i in range(2,11):
```

Although this improves its performance for 20-city problems and even for some 30-city problems, the program often gets stuck in a non-optimal rut, like in Figure 12-11.

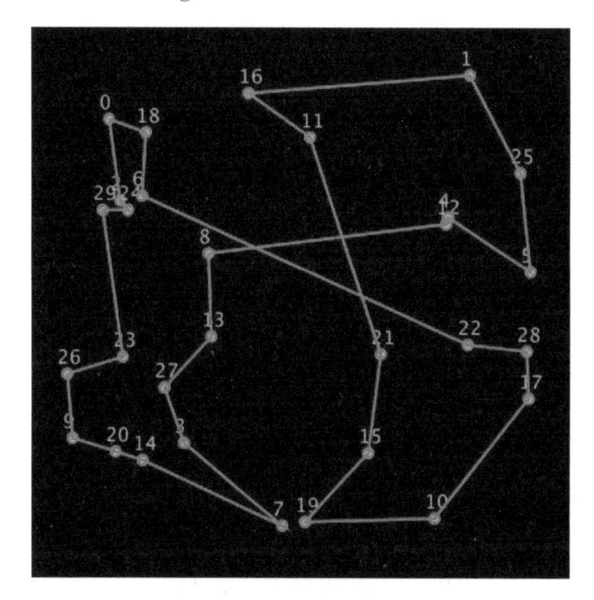

Figure 12-11: A 30-city problem stuck in a non-optimal rut

We're going to take the final step and go fully genetic. Now we won't be restricting ourselves to what we think is the best route so far. Instead, we'll have an enormous population to choose from.

We'll make a `population` list for any number of routes we want, we'll take the "fittest" ones, cross their number lists, and hopefully make an even better route! Just before the `setup()` function, after the `cities` list, add the `population` list and the constant for the number of routes, as shown in Listing 12-18.

```
cities = []
random_improvements = 0
mutated_improvements = 0
population = []
POP_N = 1000 #number of routes
```

Listing 12-18: Starting a `population` *list and a variable for population size*

We just created an empty list to put our population of routes into, and a variable for the total number of routes. In the `setup()` function, we fill the `population` list with `POP_N` routes, as shown in Listing 12-19.

```
def setup():
    global best,record_distance,first,population
    size(600,600)
    for i in range(N_CITIES):
        cities.append(City(random.randint(50,width-50),
                           random.randint(50,height-50),i))
    #put organisms in population list
    for i in range(POP_N):
        population.append(Route())
    best = random.choice(population)
    record_distance = best.calcLength()
    first = record_distance
```

Listing 12-19: Creating a population of routes

Notice we had to declare the `population` list to be a global variable. We put `POP_N` routes in the `population` list by using `for i in range(POP_N)`, and then we made a randomly chosen route the best one so far.

CROSSING OVER TO IMPROVE ROUTES

In the `draw()` function, we're going to sort the `population` list so the `Route` objects with the lowest lengths are at the beginning. We'll create a method called `crossover()` to splice the `cityNums` lists together at random. Here's what it'll do:

```
a: [6, 0, 7, 8, 2, 1, 3, 9, 4, 5]
b: [1, 0, 4, 9, 6, 2, 5, 8, 7, 3]
index: 3
c: [6, 0, 7, 1, 4, 9, 2, 5, 8, 3]
```

The "parents" are lists `a` and `b`. The index is chosen randomly: index 3. Then a list is sliced off between index 2 (`7`) and index 3 (`8`), so the child list starts `[6,0,7]`. The remaining numbers that aren't in that slice are added to the child list in the order they occur in list `b`: `[1,4,9,2,5,8,3]`. We concatenate those two lists, and that's the child list. The code for the `crossover()` method is shown in Listing 12-20.

```
    def crossover(self,partner):
        '''Splice together genes with partner's genes'''
        child = Route()
        #randomly choose slice point
        index = random.randint(1,N_CITIES - 2)
        #add numbers up to slice point
        child.cityNums = self.cityNums[:index]
        #half the time reverse them
        if random.random()<0.5:
            child.cityNums = child.cityNums[::-1]
        #list of numbers not in the slice
        notinslice = [x for x in partner.cityNums if x not in child.cityNums]
        #add the numbers not in the slice
        child.cityNums += notinslice
        return child
```

Listing 12-20: Writing the `crossover()` *method of the* `Route` *class*

The `crossover()` method requires we specify the `partner`, the other parent. The `child` route is created, and an index where the slicing will take place is chosen randomly. The child list gets the numbers in the first slice, and then half the time we reverse those numbers, for genetic diversity. We create a list of the numbers that aren't in the slice and add each one as it occurs in the other parent's (or partner's) list. Finally, concatenate those slices and return the `child` route.

In the `draw()` function, we need to check the routes in the `population` list for the shortest one. Do we need to check each one like before? Luckily, Python provides a handy `sort()` function we can use to sort the `population` list by `calcLength()`. So the first `Route` in the list will be the shortest one. The final code for the `draw()` function is shown in Listing 12-21.

```
def draw():
    global best,record_distance,population
    background(0)
    best.display()
    println(record_distance)
    #println(best.cityNums) #If you need the exact Route through the cities!
❶   population.sort(key=Route.calcLength)
    population = population[:POP_N] #limit size of population
    length1 = population[0].calcLength()
    if length1 < record_distance:
        record_distance = length1
        best = population[0]

    #do crossover on population
❷   for i in range(POP_N):
        parentA,parentB = random.sample(population,2)
        #reproduce:
        child = parentA.crossover(parentB)
        population.append(child)
```

```
    #mutateN the best in the population
❸ for i in range(3,25):
        if i < N_CITIES:
            new = best.mutateN(i)
            population.append(new)

    #mutateN random Routes in the population
❹ for i in range(3,25):
        if i < N_CITIES:
            new = random.choice(population)
            new = new.mutateN(i)
            population.append(new)
```

Listing 12-21: Writing the final draw() function

We use the sort() function at ❶, and then trim the end of the population list (the longest routes) so the list remains POP_N routes long. Then we check the first item in the population list to see if it's shorter than the best route. If so, we make it the best, like before. Next, we randomly sample two routes from the population and perform a crossover on their cityNums lists and add the resulting child route to the population ❷. At ❸, we mutate the best route, swapping 3, 4, and 5 numbers, all the way up to 24 numbers (if that's less than the number of cities in the sketch). Finally, we randomly choose routes from the population and mutate them to try to improve our distance ❹.

Now, using a population of 10,000 routes, our program can make a pretty good approximation of the optimal route through 100 cities. Figure 12-12 shows the program improving a route from an initial length of 26,000 units to under 4,000 units.

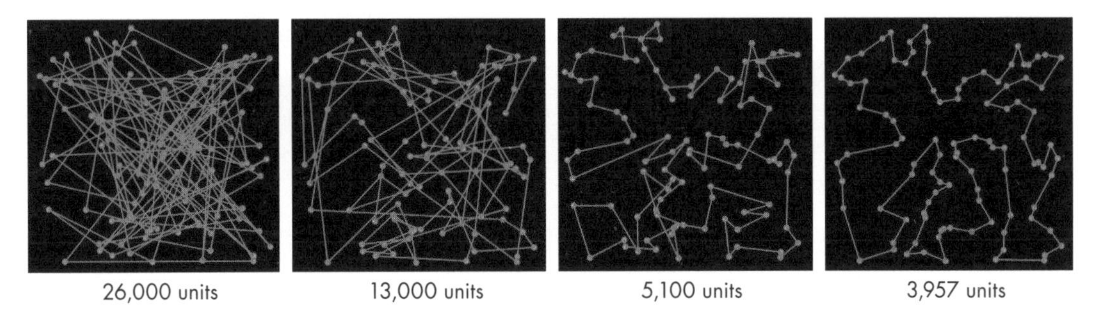

Figure 12-12: Improvements of the route through 100 cities

This took "only" a half an hour to crank through!

SUMMARY

In this chapter, we didn't just use Python to answer the types of questions you get in math class whose answers are already known. Instead, we used indirect methods (scoring a string of characters or a route through a bunch of cities) to find solutions to questions without an answer key!

To do this, we mimicked the behavior of organisms whose genes mutate, taking advantage of the fact that some mutations are more useful than others for solving the problem at hand. We knew our target phrase at the beginning of the chapter, but to figure out whether our final route was the optimal one, we had to save the city locations and run the program a few more times. This is because genetic algorithms, just like real organisms, can only work with what they start out with, and they often end up in a non-optimal rut, as you saw.

But these indirect methods are surprisingly effective and are used extensively in machine learning and industrial processes. Equations are good for expressing a very simple relationship, but many situations are not that simple. Now you have plenty of useful tools, like our "sheep and grass" model, fractals, cellular automata, and, finally, genetic algorithms, for studying and modeling very complicated systems.

INDEX

E

F

G

H

I

J

K

L

M

N

O

P

Q

R

S

T

V

W

Math Adventures with Python is set in ITC New Baskerville, TheSansMono Condensed, Dogma, and Housearama Kingpin. The book was printed and bound by Versa Printing in East Peoria, Illinois. The paper is 70# Evergreen Skyland. The book uses a layflat binding, so when open, the book lies flat and the spine doesn't crack.

RESOURCES

Visit *https://nostarch.com/mathadventures/* for resources, errata, and more information.